IF I TOLD YOU MY STORY

---✳---

RHONDA BOND

xulon
PRESS

Copyright © 2016 by Rhonda Bond

If I Told You My Story
by Rhonda Bond

Printed in the United States of America.

Edited by Xulon Press.

ISBN 9781498485388

All rights reserved solely by the author. The author guarantees all contents are original and do not infringe upon the legal rights of any other person or work. No part of this book may be reproduced in any form without the permission of the author. The views expressed in this book are not necessarily those of the publisher.

Unless otherwise indicated, Scripture quotations taken from the New King James Version (NKJV). Copyright © 1982 by Thomas Nelson, Inc. Used by permission. All rights reserved.

Scripture quotations taken from the New American Standard Bible (NASB). Copyright © 1960, 1962, 1963, 1968, 1971, 1972, 1973, 1975, 1977, 1995 by The Lockman Foundation. Used by permission. All rights reserved.

Scripture quotations taken from The Message (MSG). Copyright © 1993, 1994, 1995, 1996, 2000, 2001, 2002. Used by permission of NavPress Publishing Group. Used by permission. All rights reserved.

www.xulonpress.com

INTRODUCTION

If I told you my story
You would hear Hope that wouldn't let go
And if I told you my story
You would hear Love that never gave up
And if I told you my story
You would hear Life, but it wasn't mine

If I should speak then let it be
Of the grace that is greater than all my sin
Of when justice was served and where mercy wins
Of the kindness of Jesus that draws me in
Oh, to tell you my story is to tell of Him

Big Daddy Weave

ACKNOWLEDGMENT

The completion of this book would not have been possible without the assistance and expert advice of many people. Many thanks to all who have given me support and contributed to this project. And I thank my Mother and family for their lifelong prayers.

CONTENTS

ONE
    FAMILY 19
TWO
    A VISITOR AT THE WELL. 34
THREE
    SCHOOL 36
FOUR
    IT GETS BETTER. 41
FIVE
    CHANGES 45
SIX
    SUMMER. 50
SEVEN
    CRAIG 52
EIGHT
    DRUGS 54

NINE
1969 57
TEN
PIG PEN. 59
ELEVEN
TIME TO MOVE 62
TWELVE
HAVING MY BABY 66
THIRTEEN
ON MY OWN 66
FOURTEEN
A NEW JOB 74
FIFTEEN
BRYAN. 76
SIXTEEN
CARY. 83
SEVENTEEN
DR YA NOW 85
EIGHTTEEN
THE ABORTION CLINIC. 87
NINETEEN
SHERAH 91

Contents

TWENTY
    THE FARMHOUSE 95
TWENTY ONE
    ARKANSAS OR BUST 98
TWENTY TWO
    TONY GETS MARRIED 103
TWENTY THREE
    WINTER AT THE FARMHOUSE . . . 105
TWENTY FOUR
    ASHAMED 109
TWENTY FIVE
    I'M CLEAN 111
TWENTY SIX
    ARIZONA 115
TWENY SEVEN
    SIN CITY 118
TWENTY EIGHT
    A BRUTAL MURDER 123
TWENTY NINE
    ARIZONA DUSTSTORM 127
THIRTY
    ALONE AGAIN 131

THIRTY ONE
   A NEW JOB 135
THIRTY TWO
   RICK AND NICKY 138
THIRTY THREE
   HOME 141
THIRTY FOUR
   CLOWN SCHOOL 147
THIRTY FIVE
   AWA 153
THIRTY SIX
   DESERT STORM 160
THIRTY SEVEN
   AFRICA 170
THIRTY EIGHT
   A SCARY VISITOR 180
THIRTY NINE
   FLYING PRIVATE 190
FOURTY
   ANSWERED PRAYER 200
FOURTY ONE
   THE PRINCE 205

Contents

FOURTY TWO
- A SPECIAL GUEST............207

FOURTY THREE
- HANDS AND FEET.............211

FOURTY FOUR
- NO TURNING BACK............218

FOURTY FIVE
- LETTER TO GOD..............220

FOURTY SIX
- BOB........................225

FOURTY SEVEN
- THE DATE...................227

FOURTY EIGHT
- PAYSON.....................234

FOURTY NINE
- GOING TO THE CHAPEL AND WE'RE GONNA GET MARRIED....240

FIFTY
- THE WORD IS OUT............242

FIFTY ONE
- A DREAM COME TRUE..........245

TO THE READER

I am shockingly honest and straightforward with my approach of my story making it sometimes difficult to relive my past and disclose sensitive dealings. Often we feel life is unfair, and our circumstances and tough times in life create doubt. There have been many times in my life I wanted to give up and doubt God. But something kept tugging at my heart not to give up. The bible tells us, "consider it all joy whenever you face trials." The Lord Jesus Christ has taught me God is a detail God. If you have a passion and a desire to do something within God's will, you will achieve it. You may have hard times and feel you don't have the strength to get up or carry on. But with God ALL things are possible. It is my prayer my story will bring encouragement and reassurance to trust in the one true God.

DEDICATION

This book is dedicated to the one who cared enough to wait for me and knew I would stumble and be slow, but waited still. To the one who knew me; now I know myself. I know the pain and I know the victory. I dedicate this book to my Savior, the Lord Jesus Christ.

Come to me all who are weary and heavy laden, and I will give you rest. Take my yoke upon you and learn from me, for I am gentle and humble in heart, and you shall find rest for your souls.

—Matthew 11:28 NASB

If I Told You My Story

I also dedicate this book to my grandchildren, Riley, Cade, and Brynlee. You always make my heart smile when I am with you. There is no greater joy than doing life with you. You make me so proud to be your Nana.

Children's' children are the crown of old men and women.

—Proverbs 17:16 NASB

THE FAMILY

On December 30, 1953, I was born. I was the fourth of five children. We lived in downtown St. Louis, Missouri, on Maffit Avenue. Mama was four months pregnant with me when she started having excruciating cramps that led to bleeding. She hadn't even been to see a doctor yet. Daddy was working out of town when it happened. A dear neighbor friend helped Mama get through the night and witnessed the evidence of the loss of her baby. Mama was a tiny woman, so it came to complete surprise to her a month later at her doctor's visit when he told her she was still pregnant. I was a twin. I was born a survivor. Later in life, this would prove to be true.

Life wasn't easy for Mama and my daddy. Dwight D Eisenhower was president and had given his famous "Atoms for Peace" speech. The speech was given to try to calm the fears of the American public and enlighten them on the risks and hope of a nuclear future. Jobs were scarce. A demand driven recession occurred due to a combination of events that took place in the 1950s. The Korean War, an expected inflation period that never took place. There was increased pessimism in the economy.

Daddy and his twin brother, Alvie, would go town to town in search for jobs, leaving Mama alone with her four babies in a three room apartment.

My Daddy was a handsome charmer to the ladies. Women and alcohol he loved. Daddy was the gentlest and most giving man. That is, until his friend, alcohol, met up with him. I call it "devil juice." Because it would bring out the devil in my sweet Daddy, and Mama paid for it when he came home drunk.

The Family

We were all awaken in the middle of the night when Daddy came home all liquored up. We heard Mama's screams and her being tossed against the wall. Mama tried not to make a sound to wake her children, but we lay in bed shaking and sobbing every time.

Daddy and Uncle Alvie were the best of friends, but trouble was sure to find them when they were together. They were the youngest of thirteen children and born in the little country town of Moberly, Missouri. The twins were three months old when their daddy died of cancer. Thirteen kids, and now Grandma Bond was left alone with twin babies in 1923. That was a lot to take on, I expect. Grandma's oldest daughter was married at this time. They didn't have children yet, so Grandma asked if she would take on one of the twins. She took my daddy.

September 1, 1939 Germany invaded Poland without warning, sparking the start of WW2 and the most titanic conflict in history.

Times were tough, and the military was in need of young soldiers. As soon as Daddy and Uncle Alvie were old enough, they were drafted by US Army and headed off to boot camp at the young age of 19 years old.

At that time, the US Army allowed brothers to be together in the same platoon. Daddy fought in the Battle of the Bulge and Normandy. The United States forces incurred

The Family

their highest casualties for any operation during Battle of the Bulge War.

The battle of Normandy began June 6, 1944, also known as D-Day, when American, British, and Canadian forces landed on five beaches on the coast of France's Normandy region. It lasted from June 1944 to August 1944, resulting in liberation of Western Europe from Nazi Germany's control. The Normandy landings have been called the beginning of the end of war in Europe. More than 9,000 servicemen died in the D-Day invasion. My Daddy survived them all. Thank you, Jesus!

Daddy didn't talk much about war stories. One he did share is what took him out of the war with a Purple Heart. I don't know if this is exactly how it happened, but this is what my Daddy had told me. He was standing by a water well when this woman saw a piece of shrapnel coming at him. She ran towards him and pushed him out of the way, and took it in the heart. She died. I do believe something like this did happen, because once we were watching a war movie on TV. There was a scene something like that and my Daddy told me that was how it happened to him. I'll never forget this.

That piece of shrapnel brought my sweet Daddy and Uncle Alvie home from the War. They took both brothers out once one was injured. The war ended summer of 1945. WW2 Americans loved to smoke cigarettes. Who could blame them? Movie stars smoked, and the US President smoked. Heroes fighting the enemy smoked. And cigarettes were free for the soldiers! The tobacco company was doing well and the American economy as a whole. About a third of those cigarettes ended up in the mouths of the young GIs. Cigarettes were actually part of the GIs' rations. They would trade and sell them. Daddy smoked the Lucky Strike and Camel brands.

The Family

Coming home from the war was a real adjustment as it is for any serviceman coming home from war. Daddy and Uncle Alvie would go to church with Grandma Bond. They loved their mama and always took care of her needs. One summer Sunday morning daddy went to church with grandma. It was summer of 1946. Daddy saw this pretty little big brown eyed girl. Anna Marie. He was smitten right away and she thought this is the man I've been looking for. He fought in WW2, going to church with his mama. My daddy is a handsome smooth talker and soon won the heart of my mama. They were married August 2, 1947.

It wasn't long after they were married before Mama was pregnant with Carolyn, then Mel, and then Larry, me, and lastly, Susan. It seems Mama no sooner got her baby out of a diaper and she was pregnant again with the next child.

Marriage was tough for Mama and Daddy with work still being hard to come by without a skill or college degree. My Daddy was a hard-working man though, when he did work. He had such a gregarious personality that everyone loved my daddy. He could be so silly and always kept the kids laughing.

Uncle Alvie and my daddy were always going from town to town looking for work and leaving my mama alone with the kids.

He never wanted mama to work, and besides, it would be too difficult with all those babies.

My daddy was a good man. He was so gentle and loving, he'd give anyone in need the shirt off his back along with his last dollar. That is until he met up with his friend, Alcohol. Alcohol is not called spirits for nothing. I call it devil spirit because it sure brought out a devil spirit in my daddy. He wasn't the same person when he and alcohol hung out. My daddy got mean with the spirit of alcohol and he would come home and take it out on my mama. He had a mean streak in him from that spirit that caused us all to freeze.

We moved a lot. When I started school we lived in Troy, Missouri. That house was a two-bedroom shack down a hill off the hi-way nestled in the middle of the woods. The driveway was kind of a steep hill with gullies on each side filled with trees and brush. I have some of my fondest childhood memories from that old house. All of us kids slept in one bedroom. The girls slept together in one big bed and the boys each had their own twin bed. My poor sis, Carolyn was always forced to sleep in the middle as me and Suzi both wanted to sleep next to her. She would tell us stories at night to put us to sleep.

Summers were always so hot in that little house. We shared a small revolving fan in our bedroom. Like most little kids, when you're hot, it's uncomfortable, and trying to fall asleep even more uncomfortable. We're

The Family

flopping around on the bed trying to get comfortable. Carolyn used to say, "Lay still and quit moving around, and you won't be so hot."

This house had a basement that you had to walk outside of the house to get to it. In the basement was a pot belly stove and a wringer washer. The washer has two metal wash tubs. One for washing the clothes, the other with the rinse water. We helped with the washing by catching the clothes as mama wrung them through the wringer. Then helped with hanging the clothes on the clothes line in the yard to dry.

We took our bathes in those metal wash tubs as we didn't have a bathroom with a bathtub. Using the pot belly stove for heat in the winter. Our toilet was the old "outhouse." There were no street lights, so of course it was pitch black at night. We had a septic tank fairly close to the house, so when we needed to go potty when it was dark we would go behind the septic tank. Unless we took

If I Told You My Story

a flashlight, if we had working batteries, we would brave it and go out to the outhouse. Then there was always the fear of stepping on a snake! *Yikes!*

Summers were always busy. We had a large garden. We grew everything we could. Then mama would can everything she could using those wash tubs again. Those tubs got more use! We would go blackberry and strawberry picking. The blackberries were usually wild.

One of my most favorite Sunday meals was chicken and dumplings and blackberry pie or cobbler. I've tried copying the recipe just from memory, and my family loved it! Mama would have us kids chase down a hen, and then she would wring its neck and take it to the ol' stump and chop its head off. That ol' hen would take off running around with it's head off. It wasn't a pretty picture, but it made a wonderful Sunday dinner and lots of memories.

The Family

We had lots of chickens running around and ducks too. We all had our own pet chicken or duck or both. Then we got pigs and a goat! We started out with just a few piglets. We would get in the pig pen and play with those piglets. Yikes! that's dangerous! It was okay until they started getting big. Big Red grew huge. 'Big Red' we called him. Once Big Red broke out of his pen. Everyone was running around chasing him to get him back in. Me and Suzi were afraid, so we ran in the house. My daddy was a strong man. He got Big Red and put him back in the pen. It wasn't long after that we saw Big Red hanging from a tree. Big Red was being sold. We cried for days.

Daddy carved us wooden animals to play with. We stood them against the trees in the woods and played like we were in the jungles of Africa.

I guess we really believed we were in Africa, we had grape vines that hung from huge old trees. The most fun was swinging over this huge gulley from one side to the other. Once Carolyn and me swung together and the vine broke! Luckily it broke pretty much once we got to the other side, but we fell right next to a snake. I never saw it but she later told me about it.

We had a loving God fearing mama and her prayers were answered asking the good Lord to send out Angels to protect her crazy children. I'm sure God looked down and saw how much fun we were having and sent our Angles to protect these wild children.

The Family

We had a water well where we got our water with a hand pump. It was a deep well, but sometimes it would go dry then we were out of water. When it was dry we all got our buckets and would head into the woods where there was an ice cold bubbling stream of water. We would fill our buckets with as much water as we could. and carry back to the house, then go back and do it again until we couldn't move anymore.

It was always fun at first.

A VISITOR AT THE WELL

One summer day everyone was gone except my mama, me, and Suzi. Suzi and I were playing by the well. Then I noticed a man walking down our driveway to our house. Frightened, I grab Suzi and run into the house and told mama a bum was coming down the driveway. Back then that is what a homeless or vagrant person was named. Mama pulled us into the house and locked the door. Pretty soon this vagrant man came knocking on our door. My mama asked what he wanted through the closed door. He said, "ma'am, I'm just hungry. Can you give me some food?" Mama told him she would make him some food and he had to sit on the

A Visitor At The Well

well to eat it. She fried up some bacon and a fried egg and made him a bacon egg sandwich and gave him coffee. She gave it to him through the door. Closed and locked the door and jammed a butcher knife in the door jam. I guess that was her sign to not mess with her. I can still see that dirty old guy sitting on the well eating. We felt sorry for him and he was so nice. Knowing my mama, I'm sure she prayed for him. When he was finished with his meal, he got up and walked back up the hill and disappeared. God's word does advise us to love and serve everyone, that we never know when we may be serving an Angel of God. I believed this was one of those times.

Do not neglect to show hospitality to strangers, for by this some have entertained angels without knowing it.

Hebrews 13:2 NASB

SCHOOL

I started first grade at this old house. Mrs. Heckler was my first grade teacher. Funny I still remember her name and don't remember any other teacher's name. For some reason Mrs. Heckler did not like me. I don't know why; I was a cute sweet little girl. Was it because we were so poor? Mama always made sure we were clean with clean clothes. I had long golden brown hair and big brown eyes. Long banana curls were my usual style.

We rode the bus to school. My brother, Larry, would walk me to my classroom. I would cling to him with tears in my eyes when he said he had to leave. If I could have dropped out of school in first grade, I would have.

School

We were learning to add and subtract. Mrs. Heckler had us sit in a circle on the floor as she did addition and subtraction number flash cards. I wasn't so quick with my answers. We took turns giving the answer when Mrs. Heckler pulled out a number card. Being nervous to be on the spot, my brain would freeze and I was slow to give my answer so she made fun of me in front of the class. The kids would laugh and I wanted to cry so bad but held it in.

If I Told You My Story

Once I had to go to school being sick to my stomach, and I had diarrhea, probably due to the stress I was going through with this mean schoolteacher. Mama told me to put toilet tissue in my panties in case I had a leak. Oh my! We were having a big math test on this day. I'm sitting at my desk concentrating on my test and the little boy sitting next to me said, "Psst! Rhonda, you have paper hanging down from you skirt." Oh my gosh! I was horrified! The tissue paper had worked its way out of my panties and hanging out of my skirt to the floor. I think it was clean, thank goodness. I got a hundred percent on the math test, and Mrs. Heckler thought I cheated.

Winters were hard at this old house. This little wooden house just was not built well. There was one particular winter that was really bad. Snow was coming into the house through the cracks. It was so cold and miserable. Grandma and grandpa Bray, mama's parents, persuaded mama to bring her five

School

children and come live with them until there was a break in the storm. They lived in the suburbs of St Louis. So we packed up and moved in with grandma and grandpa.

Grandma was not a kid-friendly grandma. We were all afraid of her. She made us go to school there, which was very different and awkward. Coming from a little country school to a city school, we were probably like the Beverly Hillbillies or Duck Dynasty to these city kids.

I always say that you can always make a positive out of a negative. The positive here for me was that I was introduced to Billy Graham through Grandma's TV. Grandma had Billy Graham on TV all the time. We didn't have a TV, so it was interesting to watch. Even back then, Billy Graham's teaching was interesting and held my attention.

IT GETS BETTER

Daddy learned the trade of a plumber working with his cousin, Virgil. Virgil was a General Contractor building new homes. This was the best thing that came to my daddy in years. He was beginning to make decent money now. But then his friend, alcohol, came to visit again. One evening, Daddy came home late from work after stopping off at the local tavern. He drove a pickup truck and lost control of the wheel—or did he fall asleep at the wheel? I'm not sure, but his truck rolled off the drive into the gully. Buried amongst the trees and bushes, we thought he was dead. But God wasn't ready for Daddy to come home yet. He still had a lot to learn.

If only Daddy had pulled himself together and stayed away from alcohol. He drank all our grocery money. Mama had to step in and do something to make money. She started ironing clothes for people at the church or anyone who needed her service. I used to love hanging out with her and watching her iron clothes. She made it look so fun. She charged fifty cents a piece.

My mama was very artistic and creative. I don't know where she learned her skill, but she started making flowers out of petrified paper. It's a fine paper made from wood. She would make these gorgeous corsages for Easter, Christmas, or any occasion. She made flowers for decorations and for weddings. Mama was bringing in extra money for groceries and other essentials. At Christmas time, all of us kids would go with Mama to sell her flowers. The corsages sold for $2.50 a piece.

It Gets Better

Daddy ended up getting a job in St. Louis at Wagner Electric. He had the night shift. We were on the move again. Finally, like everyone's dream come true, Mama and Daddy were in a position to buy a house. I was in fifth or sixth grade when we moved to Harvest Acres, just outside of St. Charles, Missouri. We were really happy now. We were going to church every Sunday. That is, everyone but Daddy. We always went to church, but this church was close, and it was like family. I learned so much about the Bible and who God is through my Sunday school. Dressing up for holidays was always so fun. I loved my Easter bonnets and chiffon Christmas dresses to wear to church. I felt so pretty.

Once we even went on a family vacation to Lake of the Ozarks. I still have fond memories of that one and only vacation. We couldn't afford a hotel, so we all slept either in the station wagon or on the ground with blankets and a pillow. We cooked out at the campgrounds. I loved the smell of the outdoor cooking. We cooked hamburgers and hotdogs and played in the park.

CHANGES

I was twelve years old when Mama got a job at Moog, Ind. It was a factory job. Daddy was against it, but he started slipping into his alcohol habits again, and she knew she had to do something to keep our house together. Many Friday nights, Daddy came home drunk, and we were awaken up by Mama's cries and her being thrown against the wall. No child ever forgets this sound. I hated his alcohol breath the next day, but I still loved my sweet Daddy. He was so warm and cuddly. I knew he loved us more than anything, but he needed help, and no one knew how to help him.

Mama started getting pretty confident now that she had a job and was bringing in good money. She started listening to and getting the wrong advice from her co-workers. Pretty soon, we were going to church less and less.

One summer morning while Mama was at work and Daddy had just woken up from working all night, there was a knock on the door. "Daddy, it's the police!" Daddy came to the door, and the sheriff delivered his divorce papers. I'll never forget the look on my sweet daddy's face and how his heart was just pierced. Tears welled up in his eyes. This was

Changes

the beginning of the end. Our happy family was being torn apart. Daddy started going to church and taking the kids who would go with him, but now it was too late. Mama's heart was too cold toward him now. She wasn't interested in giving him any more chances. Daddy stayed until the divorce was actually final.

Everyone in the neighborhood knew my Mom and Dad were getting a divorce and made fun of us for that.

This divorce destroyed my daddy more than he realized it would. He really loved Mama more than anything in the world, but she was done.

Daddy moved back to Moberly. At this time, his twin, Uncle Alvie, was back in Moberly as well, living with Grandma Bond, so Daddy moved in with them as well. They were his second greatest loves in life. His family was first, and then his twin brother and his mama. Daddy just needed to know

If I Told You My Story

there was someone who wanted to help him and love him.

Daddy went from bad to worse in Moberly. We never saw him again or ever heard from him unless we made contact with him, and he was usually drunk. Wrong or right, I chose never to contact him again.

Well, now that Mama was a single mom, she worked as much as she could, even if it meant overtime shifts. That meant we had to pitch in and help out more now. Mama needed us now. I was twelve years old when I first started cooking.

Carolyn finished high school and got a job at McDonnel Douglas, where she met her husband, Russ. They married. What a wonderful husband Russ is for my sister. She was now gone from our home, and Melvin joined the Army. So now Larry, me, and Suzi were left to help our Mom with keeping up the house.

I made dinner every night, so by the time my Mom got home from work, dinner was

Changes

ready. Suzi and I cleaned the kitchen together or took turns. I loved our home and loved to keep it clean, so I was always cleaning the floors and helping out with the chores. I actually enjoyed it and knew Mama did too.

I loved my brother Melvin so much. I loved hanging out with him. It was hard for me when he went away to boot camp. When he went to Korea, and I wouldn't see him for at least a year, I really took it hard. He was so cool and good-looking. All the girls at school wanted to know me because I was Mel Bond's sister. I loved sitting next to my big bro in his car and riding through town, but now he'd gone off to the Army. He was always so protective with me and took over being the man of the house, but now he was gone too. Life was really changing now.

SUMMER

I wanted to get a summer job. My friend Valerie and I decided to try to get jobs as car hops at Dog N' Suds, a root beer, hotdog, and hamburger joint. We went to apply. We were both fifteen. We filled out our applications. Valerie went first to talk to the boss. This man was as round as he was tall, and bald, with a dark mustache. Valerie came out from her interview, and I was next. After talking with me, he asked how old I was. I couldn't lie and told him the truth: "Fifteen." Valerie lied and said she was sixteen. I got hired.

This man was very protective of me. He told me, "Don't talk to those boys coming in. They are only going to get you in trouble." I didn't last long at Dog N' Suds. Larry would drop me off and pick me up on his motorcycle, and that got old.

As a family, we quit going to church altogether now. Mama was working two jobs at times. Larry started going out with a girl down the street, who he eventually married, so he was hardly ever home. Mama was single and living the single life now. She was always gone on Friday nights and looking for the love she never found.

We kids befriended a young neighborhood couple, Phil and Sam. Sam was in her mid-twenties, young and pretty, with long blonde hair, sky blue eyes, and long tanned legs. All the neighborhood kids loved hanging out with Phil and Sam. They let us do anything we wanted. The kids would join them in drinking beer and smoking cigarettes. Larry and Suzi hung out there more than I did.

CRAIG

I was in the seventh grade when I first saw Craig touring the seventh grade school with his sixth grade class. By January, 1968 in my eighth grade year, Craig and I would be dating. It was a weekend, and all the insubordinate kids were there. The pond was frozen over, and we were ice skating. They all brought alcohol and food. Somehow, underage kids will always find a way to get alcohol. I started drinking a little. I never liked beer and didn't really even care for alcohol much, but I drank because all my friends did. I was a tiny girl, so it didn't take much before I was buzzed.

That's the night when Craig and I first met, and our eyes locked. It was love at first

Craig

sight—puppy love, maybe. He had tight curly brownish blonde hair and green eyes. Craig had dark skin tone and a mustache. That was it! We were dating now. Craig was such a fun guy, with the most infectious laugh. Everyone loved Craig. He was the life of the party, and he could sing. Craig would strum on the guitar and sing. He had a great voice. We were so in love. We knew we would be married some day.

Craig's best friend was Ricky. Ricky had an older brother, Bobby, and younger sister, Patty, who became one of my best friends. Ricky's family was quite wealthy. It was a fun family. They would take all of us kids to their cabin at the lake in the summer. It was a blast. Swimming, BBQ's—life has never been so good. We sat around the evening campfire talking until wee hours of the night telling Rick's parents all we needed in life was peace and love. We were a population of the flower power, peace and love. It was groovy.

DRUGS

There seemed to be no rules at home. I was a teenage free bird. In the summer of 1968, Craig and I joined the hippie movement. I wore halter tops and torn bell-bottom jeans, and was barefoot with straight long brown hair. We were hippies. We didn't buy our jeans torn and tattered; we just couldn't afford to replace them, so we wore them until they were worn and patches sewn on the torn holes would no longer hold up, or there were just too many patches. We were thrilled when we could afford to get a new pair of jeans. This was all about love and peace. We would go to the mall in St. Louis, Northwest Plaza, and hang out around the fountain and panhandle money for fun, though sometimes we really needed money to eat.

I would sneak out of my bedroom window and not come back home for days in the summer, and even sometimes during school on weekends. This is where I was introduced to drugs. I was never interested in doing drugs, but all my friends did them. I was just the free-bird little girl who was learning to color outside the lines of life. Like a fish out of water, I was so lost and didn't know it. My schooling suffered. It didn't seem to matter how I did in school, so I didn't care, to a degree. I got through and passed every class, but I sure didn't try hard enough to learn. I didn't realize that someday I would need education and how important it was.

Some of my guy friends thought it was fun to steal cars. I would drive the car while some of them snuggled in the back seat high on marijuana, alcohol, or other hallucinogens. I didn't even have my driver's license when I

If I Told You My Story

was driving. God surely was watching over me. He wasn't ready to let my life come to an end. I'm not sure what those guys did with the cars afterward. Did they take them back? I don't know. As far as I know, they never got caught. My mother would be upset with me for leaving for days and not knowing where I was, but I guess she just didn't know what to do with me.

Our friends nicknamed Craig "Lamby" because of his long locks of curly hair. We were at a party one night. Our friend Bryan was doing LSD. He was so high he climbed on top of the roof of the house and said he was Superman. He really thought he could fly. Bryan died that night.

Craig and I hitchhiked everywhere. Sometimes I hitched alone. Surely God sent his angels out to watch over me. He knew the plan in my life and that I would eventually turn around and come home. Why did it take me so long? My heart grieves today at this question.

1969

It was August 15, 1969 when Woodstock ignited in NY. Craig wanted me to hitchhike with him there. I didn't go, so he and Ricky left without me.

It was summer 1969, and Ricky's parents were out of town. All the kids stayed home. It was time to party. A huge party broke out at Ricky's house for the whole weekend. There was music blaring, alcohol, drugs, and sex all over the house. One of my best friends, Dennis, came to the house with several guys I had never met before. Dennis was tall and muscular, with beautiful blue eyes. He was in the Army and was AWOL.

Kathy had broken up with Dennis, and he was heartbroken. He just couldn't pull himself together. Dennis came to me privately and told me he was going to go with these guys he came to the party with and shoot up heroin. I begged him not to, that no girl or anyone is not worth that. My friend, Valerie, over heard the conversation and gave him a piece of her mind and told him how stupid he was if he did that. My tears and persuasion didn't help. Dennis turned on his heel and headed out with the guys. Why do we do this to ourselves? If someone chooses to leave, let them go. Give yourself a day to weep, and then take your first step forward.

PIG PEN

Very early the next morning, the same guys came back with Dennis. Only this time, they carried Dennis' body in the house. Dennis lay on the floor in the living room with his shirt pulled off. His jeans just hung on him. The guys said he overdosed. Apparently, we found out later, the heroin was laced with too much strychnine, and my Dennis couldn't pull out of it. He struggled for a breath, and his stomach sank to the floor, begging for another breath. Poooofh! He would let it out and quickly gasp for another breath again. It was horrific! I cried and cried, squeezing his hand and saying, "Please wake up, Dennis."

If I Told You My Story

Not knowing what to do, the guys phoned the apparent drug dealer and were told to soak him in an ice cold tub of water. The guys carefully and quickly carried Dennis's limp body upstairs to the bathroom. They dumped ice cubes in the cold bath water. This was supposed to shock his vitals into functioning, but it did nothing. The guys made the girls leave so we didn't have to see what happened. They probably also wanted us to get out of the way. It wasn't working for Dennis, so the guys said they were going to take Dennis to the emergency room and would let us know how he was. That was the last time I saw my Dennis alive. We all cleaned the house and ended the party.

Three days later, the St. Charles police came pounding on the doors of Ricky and Craig's homes at three o'clock in the morning. They took the boys down to the police station. The police told Ricky and Craig that the guys who gave Dennis the heroin never took him to

Pig Pen

the hospital. They dumped his body in a pig pen! My dear friend, who was fighting for his life, died in a pig pen. We were beyond tears.

We were all there for Dennis's funeral, with the exception of the guys who drugged and killed our friend. Beautiful nineteen-year-old Dennis lay in this casket. A killer drug took the life of this young boy way too soon. I'll never forget Dennis's sad eyes the last time we talked. The next time I saw him, death was waiting for him.

TIME TO MOVE

It wasn't long before mama realized she couldn't afford the house anymore, so she sold it and bought a mobile home. Although it was nice, it just wasn't the same. We moved way out in the country, to O'Fallon. It was country in 1969, anyway. Suzi did not like O'Fallon at all. She wanted to stay in town and finish school so she could be with her friends. She moved in with Carolyn and Russ so she could do so.

Craig and I were always together when we could be. Either he hitchhiked to see me, or I hitched into St. Charles to be with him. Many times I spent the night at his house. His parents had recently divorced. His mother was very angry about the divorce. His dad was a cheater and found another woman.

It was the spring of 1969 now, and I was pregnant. I was a tiny girl, so it wasn't long before my baby bump showed up. Craig and I were excited. We were planning our wedding. After mama found out about it, she was not happy at all. I come from generations of a very religious Pentecostal family. This was not going to go over well with the family. Like please, this Christian family has never done anything wrong? Seriously, I'm the only sinner?

Their sin just wasn't exposed like mine was, I'm sure. Mama was not going to allow her sixteen-year-old pregnant daughter to marry this fifteen-year-old boy. Even though she liked Craig, marriage was no option. My friends gave me a baby shower. I used to organize those baby clothes and toys different every day and just imagine my baby in them.

If I Told You My Story

I was not allowed to go back to high school pregnant. I was so brokenhearted.

Mama finally found the love of her life, Hank. It was a blind date, but love at first sight. Now she had to tell him she has a pregnant teenage daughter, and she's going to be a grandmother. It wasn't long before Mom and Hank married. Everyone loved Hank. He was easy to like. We were all very happy for Mama. Hank moved into the trailer with us. Hank was divorced, with three children.

Craig would hitchhike 20 miles to O'Fallon to see me as often as he could during the day when I was alone. Now that mama was married, there were new rules, and she didn't want me to see Craig at all now. He was barred from the house. Craig said, "Let's go to California." When I asked how we would get there, he said we would hitchhike. I just couldn't do it.

Craig, so broken, now joined the Army.

Being pregnant in school was not accepted yet. Shoot, we still had to wear dresses to

Time To Move

school. I was kicked out of high school because I was pregnant. The next year, the schools started allowing pregnant girls to finish school and also allowed girls to wear any kind of jeans to school. Evidently pregnancy became an epidemic after 1969, so it was allowed in school? A lot of changes took place after I left. I remembered being thrilled that we could start wearing jeans to school, even though they had to be white jeans.

HAVING MY BABY

Life was not easy for me now. Mama and Hank wanted to send me away to have the child, and they would raise my baby as their own. I wouldn't have anything to do with that plan. This was my baby. I was going to have my baby and raise this baby the best I could on my own. I was sure that I would either get married soon or get a great job to support us.

Anytime relatives would drop over, I had to hide. Mama made me hide in the closet in my bedroom until they left. I was in the closet almost an hour once.

I had a gentle, really nice doctor. Dr. Reese never talked much, but I felt comfortable with him. He gave me a pink book that described every month in detail describing what was going on with my pregnancy and the growth of my baby. I read that book daily, like it was the Bible. Pregnancy alone was hard on me. Because I was living so far away, few friends were few came out to see me.

It was December 18, 1970 when the labor pains started. It woke me up at about 5am, but it was only about 15 minutes apart at this time. I was really craving Entemmans chocolate donuts, so Mama ran out to get me some. By that night, labor pains started getting closer, so we were off to the hospital. Dr. Reese was not on call, so his partner, Dr. Duncan, took his place to deliver my baby. He was a tall lanky man with glasses, long arms, and big hands. He was kind of scary to me.

The nurses were awesome. My nurse was very gentle and motherly with me as I lay there in great pain. I didn't scream like I heard others in the rooms next door. My nurse would massage my lower back, where the pain was the greatest. At 3:30am on December 19, my baby was ready to be held in my arms.

Dr. Reese never explained to me the steps or what procedures would take place in the delivery room. I didn't know what to expect and was frightened going into the delivery room. I only knew I was in grave pain and could hear the nurse scream, "I see the head!" I was a scared little girl. Pretty soon, the doctor pinned my arms and legs down in a brace so I couldn't move. I had no idea why they were doing this to me. Then, while trying to hold my wiggling head still, they covered my nose and mouth with sleeping gas. I was terrified and felt like they were trying to kill me!

I woke up in another room all alone with a nurse roughly pushing on my belly. *Ouch!* That hurt! My first reaction was to push her hands away. Apparently I scratched her, and she slapped my hands and said, "Don't you be scratching me." She was pushing the afterbirth out. Goodness, do they still do it that way?

Having My Baby

I was in the hospital for five days after delivery. I couldn't walk or go to the bathroom for a couple of days, so they were keeping a watch on me. I only weighed 95 pounds when I got pregnant and now I weighed 136. Maybe that had something to do with my difficult pregnancy. My only visitors allowed were my mama and brother Mel. When they brought me my baby and put him in my arms, I didn't know I could love like this. I held my 7lb 11oz baby boy and forgot all about the pain I had just gone through. I smiled and kissed my baby, with tears streaming down my face.

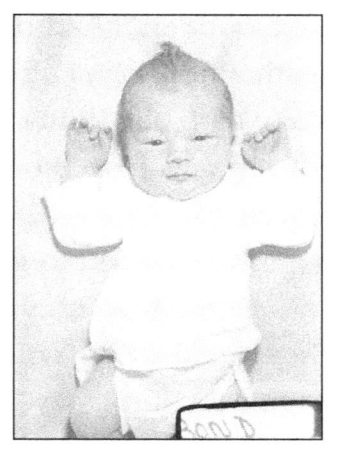

Craig and I had already decided to name our baby boy Bryan after our friend Bryan, who died on LSD. Lee was going to be after my daddy.

If I Told You My Story

Bryan Lee. Bryan was beautiful! He had black, thick hair and beautiful brown eyes, with reddish skin. He looked like a little Indian papoose.

Motherhood took some getting used to. I wanted to nurse my baby, but Mama was against it. She said people don't do that anymore. So I bottle fed my baby. I cried a lot. All of a sudden, I had no friends and felt all alone with a newborn baby that I didn't know how to take care of. I didn't know how to raise a child, but I knew I loved being a mama. That's a God given gift.

Suzi decided to get me out of the house for an evening, so she took me to a concert with her friends. She took me to see Led Zeppelin. It was wild and crazy. I didn't care for the music. The people all seemed high. We were walking through the stadium through the crowds of people when a long-haired guy walking past me grabbed my breast and squeezed and then just laughed. I was horrified in disbelief and cried.

ON MY OWN

I was ready to leave Mom and Hank now and be on my own. Mel and Carolyn talked to Daddy and persuaded him to move down to St. Charles from Moberly to help me out. He did. Daddy rented an apartment in downtown St. Charles that used to be old military barracks. We lived there a while, and Daddy started his plumbing business up again. I wasn't working yet, so I kept our apartment clean and cooked for Daddy.

Then we decided to move to a nicer apartment in town. We moved into an old red brick two-story house on Jefferson street. It was nice. I started getting out more and meeting friends. Daddy would babysit. He and Bryan had so much fun together. Bryan loved his Papa, and Daddy was such a jovial Papa. I got a job at a restaurant, Noah's Ark. I rode my bicycle or walked to work. It was a couple miles away. I had a little red wagon I would pull Bryan in to go to the market.

It was a lazy, hot summer Sunday afternoon. The apartment above me was for rent. I was home alone, and Bryan was napping, so I took a little nap on the couch myself. There was a knock on my front door that woke me. I went to the door, and a young guy stood there, inquiring about the apartment upstairs. I let him know I didn't have any information about it, and the renters were a law office next door. He reminded me it was Sunday, and they were closed. So I said, "Well, hey, you can get to the apartment through my kitchen. I can show it to you." So I let this stranger into my house and opened the kitchen door that led into the little dark hallway that led up the dark staircase to the apartment upstairs. It was nice of me, but not a good idea. Of course it was locked! I turn around to tell him, "Oops! It's locked. Sorry." He started grabbing me. I kicked and pushed him, screaming at him to get out of my house. "How dare you, when I'm trying to do something so kind for you?!" He went running out of my house. Seconds later, my daddy came driving up. He saw I was terrified and crying. It's a good thing that scoundrel was gone, because I know my daddy would have killed him!

A NEW JOB

A new service station was opening in town. It was the first of its kind in St. Charles. It had a car wash attached. If you filled up your car with a tank of gas, you got a free car wash. I was the first female to be hired. Soon, I got my girlfriends jobs there. Cars were lined up in the street to get a full tank of gas and a free car wash where a pretty young girl in short shorts would fill your car up with gasoline as well as check the air in the tires and the oil. Men always questioned and wondered if this young girl knew what she was doing, and would ask if we really knew how to do this.

A New Job

It wasn't long before Daddy met up with his friend, alcohol, again. Soon Daddy wanted to leave and head back to Moberly, so Suzi moved in with me. Together, we were able to make the rent. Suzi was going to nursing school and would babysit Bryan all the time. Suzi took Bryan a lot to hang out with her friend Melissa and Melissa's little boy, so I was always free.

Then I met Jimmy. Jimmy was a tall, handsome guy. He had dark hair and blue eyes. He fell head over heels for me. He was fun. Soon I found out Jimmy was a user—a heroin user. Jimmy made me promise I would stay away from drugs, and especially heroin. I had no desire to do drugs. I entertained marijuana on occasion with Suzi and her friends, and did some speed. That was about it, and it was short-lived. I never liked not feeling in control. Jimmy was funny, and we always had a lot of fun, but I soon grew tired of him, so Jimmy and I didn't last.

BRYAN

Bryan was two years old now. I would pull Bryan around town in that little red wagon, or I propped him up on my bicycle. I didn't have a baby seat, so I sat Bryan on the bar of my guys' bicycle. This just gives me chills to think of my baby sitting on that bar, riding around town. I can hardly stand it! Needless to say, we had an accident. Bryan, with his inquisitive little mind always at work, was looking down at the spokes. I'm sure he wondered what would happen if he put his little foot in there. *Snap!* He found out. We crashed, and my baby's foot was broken.

A nearby neighbor saw it all happen. An older grandma and grandpa-like couple came running out and saw the blood and disfigured little foot, and Bryan screaming. Without a second thought, the man lifted Bryan up, put us both in the car with his wife, and drove as fast as he possibly could to the emergency room, honking the horn all the way.

They took my baby in right away to sedate him and get him out of some pain. I finally was allowed to come in and sit with my baby until they are ready for surgery. Bryan looked up at me and said with a big grin on his face, "Mommy, your nose is big."

Bryan came out of surgery with a cast on his leg. He sure learned to get around really well with that cast. It was years before my little Bry ever was told he needed to pick up his toys and put them away again. This baby was eating out of the palm of his mama's hand now.

I was catching up on lost time and partying with newfound friends. Bryan, Suzi, and I were still living at the brick house on Jefferson Street. Suzi took Bryan everywhere with her, so I took advantage of that freedom and went off partying. Mike and Sharon were married and were having a party. These new friends were all older than me. The husbands all had Harley Davidson bikes. This is where I met Cary. Cary was ten years older than me.

Cary had a really cool Harley chopper. He was divorced, with a five-year-old son, Cary. Cary had been in the Army and fought in Vietnam. He was married before he left for Vietnam, and his wife, Shirley, became pregnant before he left. When Cary came home four years later, Shirley met him with divorce papers. I'm sure that was a double whammy to be faced with after coming home from a nasty war.

Cary and I began dating. I was his girl. Cary soon got me a job where he worked in a factory making chorine tablets for swimming pools. It was a good paying job, but the chlorine was terrible to breathe and terrible for your skin. We wore lab coats and masks. Every night, my hair smelled terrible, and even if I covered my hair, it still smelled terrible. Cary would pick me up and take me to work every day. We got to be serious, so I quit working there and started working downtown at a clerk's office where people came in to pay their car tags and taxes.

I was now living with Cary. I was so in love, or so I thought I was. I think I was feeling the need for an older man to replace my daddy and take care of me. I didn't find that in Cary, but it was too late. He left me in tears by cheating on me all the time. What a loving God; even in the midst of all my sin, he still had a hand of protection over me. I was never infected from diseases Cary brought home. When he was infected, at least, he was thoughtful enough to stay away from me. At this time, my sister, Susan, was working at the St. Charles hospital in the emergency room. Later, she told me Cary used to come in all the time to be treated for his infectious disease. Of course, this caused her to have even more of a reason not to care for him.

We were always doing BBQ or something with our biker friends and taking day or weekend trips somewhere. About fifteen

Harley Davidson bikes zoomed down the freeway. We would go as fast as we could. Cary loved to go fast. With me on the back, he would clock in at 100 miles per hour! God was with me.

In the summer of 1974, there was the largest rock festival other than Woodstock in Sedalia, Missouri. Three days of "sex, drugs and rock 'n' roll." It was a mini Woodstock and attended by fifty thousand to one hundred fifty thousand people. Aerosmith, Bob Seger, the Eagles, and Lynyrd Skynyrd played, just to name a few of the acts.

Sedalia had many concerts from Wolfman Jack, and I attended many of them. The highlight was when Glen Frey of the Eagles dedicated "Already Gone" to Richard Nixon, who resigned from presidency three years later. There wasn't an Ozark Music Festival quite like this one again.

Cary and I rode his Harley to the festival along with our friends. We slept out next to our bikes on sleeping bags and woke up to the bright morning sun. I was always forgetting to take my birth control pills. It wasn't long before I was pregnant with Cary's baby.

CARY

Cary was not ready for marriage again or to start a new family. He loved lots of girls and much younger girls than me. He soon found a new, younger girlfriend, PJ. I was booted out: pregnant and with no place to go. My family and friends all had their own lives to deal with. Everyone was happy and didn't want to welcome a young troubled, husbandless pregnant girl with a toddler into their homes. I never asked for help unless I was really desperate. I was a survivor, and I would find a way to get through this, too.

I did live with Carolyn and Russ for about a week, and knew it was time to leave. With no other option, Bryan and I lived in my 1963

Chevy until I could find us a home. I got a job at a pizza sandwich shop. I put Bryan in day care and I went to work all day five days a week. Once I pulled into a gas station to use the restroom and clean us up a bit. I forgot to lock the door and a lady opened the door on us as I was washing Bryan. Furious, she slammed the door closed and yelled out to her husband, "Some hippie is in there taking a bath!" Me and Bryan just laughed.

DR. YANOW

I needed to find an OB doctor now. I found a doctor in downtown St. Louis, Dr. Yanow. On my first visit, I was checked and it was confirmed: I was pregnant all right. Dr. Yanow was not going to be this sweet feel sorry for you doctor. He laid right into me scolding me like a Father would. I tried to hold back the tears streaming down my face. I needed this kind of Father growing up, where was he when I needed him? Dr. Yanow didn't offer me get a tissue, I had to get up and get it myself.

When I got to my car parked on the side of the street, I was leaning on the steering wheel crying when a young guy walked by hearing me cry. He reached in and touched my arm and said, "sweetie, don't cry until you really have something to cry about." I looked up at him with my wet face and red eyes and smiled. That was the last time Dr. Yanow talked to me like that. After that it was like I was his daughter. He was so protective over me like a daddy.

THE ABORTION CLINIC

I found me and Bryan an apartment in old town St Charles on Main Street. A two-bedroom duplex apartment. It didn't have an air conditioner so a friend found me one that fit in the window. It wasn't great, but it helped a lot. Cary agreed to pay my rent while I was pregnant.

I decided to check out Dental Assistant school. I went to a school down town St Louis. I'm just approaching my third month of pregnancy. I talked to the director of the school, Marc. A very handsome Scandinavian looking guy. Tall, long shoulder length very blonde hair with a big mustache and big blue eyes. I explained my situation to Marc. Marc suggested I have an abortion before starting school. He said he would take me to the abortion clinic. He made it make sense. I have a child already and no husband. It would be very difficult to raise two children alone and who would want to get involved romantically with a girl like that. So I'm off to the abortion clinic with Marc.

Arriving at the clinic after filling out my paper work, I'm called into the office with the lady that would explain what to do and how to get started. Marc went in with me. Marc and I sit across the large desk from Darla, the manager of the abortion clinic. With her short curly dark hair and big round glasses, Darla is probably forty years old. A pushy and aggressive personality, Darla begins explaining to me that I'm doing the right thing. "it's just a piece of tissue. Just a glob of tissue. It's nothing right now." That just put a dagger in my heart.

I had a baby already and everyday I would read in the book Dr. Reese gave me about how much my baby grew everyday and the changes every month, and what my baby looked like from the time of conception. Darla may have had bypassed me with her college education, but I wasn't stupid. I didn't like Darla almost from the moment I met her.

I knew by 3–4 weeks my baby had a heart beat and by 21 days it is pumping blood through a closed circulatory system, blood whose type is different from that of mine. I was furiously disgusted with Darla's pompous arrogant approach to me.

I don't even remember how it was left other than Marc and I came out of there and I told him I was not going to have an abortion. Another God moment. Had Darla been more professional and just explained the procedure and what to expect, chances are I would have had the abortion. But she spoke down to me telling me this inside me was nothing but a

If I Told You My Story

piece of glob or flesh. I believe the Holy Spirit exposed to me that very moment, that there was a baby growing in my body more than just a piece of flesh. This baby was beginning to already have a heartbeat. That opened my eyes to what I was about to do and it terrified me.

Now I wanted the best for my baby. I considered adoption. I got paper work started to adopt out my baby after birth. I informed Dr. Yanow of my decision. I 'thought' it was my decision anyway at the time. I knew my baby would have a better life than what I could give. A single young girl with a five and half year old already. I thought, I should have done this for Bryan too. My heart broke in a million pieces.

SHERAH

I continued working at the pizza place until it was time to start school. I was about five months pregnant when I went out to my car and saw I had a flat tire. I always had a good spare in my trunk. With no other option, I knew I had to change my flat tire so I could get to work, and I did. At this point I had little contact with Cary. He never stopped by to see how I was or see if I needed to anything from the market. He had a new girl in his life and didn't want to be bothered. He just sent me a rent check every month. I would see him around town with his girlfriend and would go home and cry.

If I Told You My Story

Dental Assistant school was one of the best things I ever did in my life. Why didn't I have a mentor? I know I would have done things a lot differently than the path I took. I was a fish out of water. I colored outside of the lines. No direction and no one to encourage me to take another path. Did no one care? God did but I never gave him a chance. This is where I learned the faith I have today I believe. God is a gentleman. He is going to allow us to make our own mistakes and he is not going to force his son on any of us. I'm sure he put people in my path that I just don't remember, but had I have listened, things might have been different for me. God speaks to us with that O' still small voice. If only we'll be quiet long enough to listen and take heed to it.

Be still, and know that I am God ...
—Psalms 46:10 NKJV

Dr. Yanow was very proud of me for being proactive to improve my life by starting Dental Assistant school. At about six month pregnancy Dr. Yanow was listening to my baby's heartbeat. "Beating fast, just like a little girl's should." I was so excited! I asked, "I'm having a girl?!"

I chose to have my baby the Lamaze method. I let Cary know I was taking classes and he wanted to participate so he went with me. I was still in love with Cary and had hopes he would want me back if I included him in the childbirth.

It was late afternoon when I felt the labor pains come on. Cary was still working about 20 miles away. My friend Debby phoned his work and he was there in a flash! Cary was very excited. He always wanted to have a little girl. We were advised to leave for the hospital when pains were about 10 minutes apart. We arrived at the hospital about 8pm. We were there all night long. About 5:30 in the

If I Told You My Story

morning I was ready to give birth. Focusing on my breathing techniques for labor pain and trying to relax was not easy.

Cary must have forgotten his role. The nurses ran out to get for Dr. Yanow, he came in to see me. Giving orders to roll me into the deliver room, excited, Cary asked Dr. Yanow where he should go. Oh my! Dr. Yanow really did want to tell him where he should go. He did give him a piece of his mind and told him he would "not" be in the delivery room. Cary was furious and I didn't care.

I watched my little bundle of joy being birthed. Coming out with a cry. Dr. Yanow whispered in my ear, "Did you decide what you are going to do?" Meaning am I still going to adopt my baby away. I looked up at his contorted face and said, "I can't give my baby away." With a smile he said, "Good choice." Life was not going to be easy and I knew that, but I don't give up.

THE FARMHOUSE

I reconnected with a high school friend, Tony. Tony was divorced with a couple kids. We decided to rent a place together and help each other out. Tony had two little girls. Cher and Kimberly. They were both about the same ages as my kids so this was going to work out great. Tony and I rented a two story farm house. I took the upstairs two bedrooms and Tony had the bedrooms downstairs. Tony's family lived close by. They were all very close and Tony spent a lot of time with her family. I got a job working for a Dentist. Life was going to be good now.

Tony and I started attending church at the 'Sheep Shed.' Bryan loved it. He was a little preacher man. Sherah and Kimberly were the same age, six months. The charismatic movement was just starting in 1975. This church held about 2000 people. We loved it! There was so much real love there. We sang songs out of the Bible. The Pastor was on fire for God. I got baptized there. This is where I first really met Jesus Christ as my friend. I never

The Farmhouse

knew Jesus like this before. I felt like his presence was with me everywhere I went. I got so much encouragement from reading God's word daily. Suzi started going to church with me too. We met a lot of fun Hippie Christian friends. Down to earth and real. I loved that. That was a really fun time in my life. We did weekly Bible studies at one of the couple's house on Main Street down town St Charles. We were always having a pot luck dinner at someone's house.

ARKANSAS OR BUST

Some of the guys we hung out with found a hot deal in Little Rock, Arkansas, so a group of about 10–15 guys and gals headed to Arkansas. They had their own little commune living on top of the mountain in Eureka Springs, Arkansas. Growing their own food living a vegetarian hippie life. Of course God was part of their life too. So they thought. We have to always be on the lookout as Satan is always on the prowl searching for when we are most weak. He's a wolf in sheep's clothing. And so he was.

Be sober, be vigilant; because your adversary the devil, as a roaring lion, walketh about, seeking whom he may devour.

1 Peter 5:8 NKJV

Missing my friends I drove up to Arkansas for a couple of weeks to see what it was all about that was supposed to be so awesome and visit my friends. I put Bryan and Sherah in the car and off we went. Bryan was about five and half years old now, Sherah a baby. Driving through the Ozark Mountains of Arkansas. The Ozark Mountains of Arkansas are the largest mountain range between the Appalachian Mountains in the East and the Rocky Mountains in the West. Incredible majestic and beautiful mountain ranges.

I was driving through those mountains in the middle of a huge rainstorm. It was so heavy I could barely see six feet in front of me. It was so scary. No seat belts in my car. Both my babies sleeping. Sherah about three months old sleeping on the front seat and Bryan five and half years old sleeping in the back seat. That alone is terrifying in today's world of safety and no seat belts or baby seat. I was terrified driving through this storm. I pulled over as safely as I could and prayed asking for God's protection for me and my little family.

Finally the rain let up. I stopped at the first Holiday Inn. It's rainy still but not a down pour now. I have one baby in my arms and another little hand gripping my hand. I go into the lobby to get a room. I was given my key and now I was going to put my babies in bed and go get my luggage. My car keys are missing! I locked myself out of my car. Tears beginning to well up in my eyes. There was a guy standing in the corner watching the whole transaction. He stepped up and said, "ma'am, I'll get a clothes hanger and get your keys out for you. Scared, I didn't know what his intentions were, so I said, "no thank you,

I'll deal with it in the morning." I left and went to my room and locked the door. About twenty minutes later there was a knock on the door. It was that man. He said, " ma'am, I got your keys for you." I could see them dangling through the peep hole. I got the keys, thanked him and locked the door. God was covering me and my babies.

Eureka Springs was the most beautiful place I'd ever seen in my life! The mountains were majestic. My friends made a wise choice to move here.

We had apple pancakes for breakfast, only organic foods we ate. Everyone was helpful and got along, it was so much fun and it was the family I desired and missed. This was heaven on earth. I thought I could live here too once I got myself enough money together to come back.

My two weeks went by fast and I was on my way home. Suzi had gone up with the initial group and was living up there now. I missed her. We were always so close and did everything together. She was always there for me.

It wasn't long after my departure and Suzi and some others were on their way back. This little "Christian" hippie commune had turned into a marijuana farm. They planted marijuana plants in the backyard. So this was the "hot deal!" The country people at the bottom of the mountain was sure those hippies were up to no good on that mountain so they got the law enforcement to check it out. They found the marijuana plants and were ready to prosecute.

The guy most involved that pretty much started this commune, Rick seemed to be a rock solid man of God. He was smart and seemed to have it together. Rick fell away from Jesus, and wasn't strong enough to carry the repercussion of his actions and took his life. Devastating for all that knew and loved Rick. The little commune ended.

TONY GETS MARRIED

Living with Tony was so much fun. We were like family. We always had fun holidays with the kids. We had a big oak tree in the backyard with a tire swing that Bryan just loved. We watched each others babies while the other did errands. This old farm house sat back off the highway. We loved it here. We even painted the inside and outside ourselves! We loved this old house.

I was in the shower one evening and heard something at the window screen. I ignored it. The next day I'm doing my Spring cleaning, yep windows, baseboards, all of it. I like to know every area of my house is clean. I pull back the curtain in the bathroom window to

If I Told You My Story

clean and the screen is cut out! Another hand of protection from God. He did not allow that intruder to enter our home.

Tony met a guy from church and now they were planning to marry, Paul was a Chiropractor. He and Tony were planning to move way out to another small town hours away. I knew I probably would never see her again, or it would be a very long time before I would. I helped her plan her Spring wedding. I was so happy for Tony and Paul. It was sweet to see that Tony was so happy to have found her love.

WINTER AT THE FARMHOUSE

Winter was tough at this old house. It was a hard Missouri winter. Deep snow. This house required oil heating. Me and Tony could only afford a quarter of a tank of oil at a time. That was about $45 each. It got so cold we closed off upstairs so me and my babies slept in the living room down stairs. It got to where we needed to get more oil every two weeks or less. Tony couldn't afford it anymore so she moved in with her parents until the storm up let. I kept closing off rooms to provide more heat for me and my little family.

Now I was the sole provider for the oil heat bill. I was turning on the small little oven to provide heat for short periods of a time. Not

If I Told You My Story

a wise thing to do, but we were freezing. It was chilling! Not only temperature wise, I was always so scared going to bed every night. It was so dark. The wind would blow and the branches on the big oak trees would scratch across the side of the house or window. I was sure someone was trying to break in.

It got so cold I finally I asked my Mom and Hank if we could come stay for the week end. When I came back, even though I left the light over the fish tank on to provide heat for the fish, it wasn't enough heat. The fish froze to death. All my plants died too.

After a big snow storm one night my car was buried deep into the snow. I called my brother, Melvin to see if he could help me dig the snow out so I could get to work. I was going to be late for work. I drove Sherah to the babysitter's house, Coco. Coco loved Sherah. Coco had Sherah from before breakfast to after dinner five days a week. She really wanted to adopt Sherah and relive my burden of being a single mama. There is no way I would ever give my babies away. Maybe that was selfish, but I loved my babies and I was going to make it work out. I dropped Bryan off at day care, slipping and sliding down the snowy highway to get to work.

We had three patients scheduled that day. Dr. Hendrich called me into his office and closed the door. He never really liked me. I don't know why. I always felt it was because I wore a cross and he knew I was Christian. Who knows, but he fired me. He said I was late and he couldn't depend on me. This was the only time I was ever late.

I was twenty-one years old and single with two babies, so I needed to find another job ASAP. I searched the want ads for Dental Assistant but they were scarce. I did go on a couple of interviews, but was never hired, or I didn't want to accept the particular job.

I had a guy friend in St. Charles, Rich. I met Rich through Cary. He and I would visit on occasion. Rich and I were just buddies. He had given me a key to his apartment so I could hang out there if I was in town. He had a Harley too. We would go ride now and again.

ASHAMED

I stopped at Rich's house after interviewing for a dental job. Rich's apartment was across the street from a local Laundromat. I was dressed up, with my long brown hair pulled up on the sides and no bangs. I was wearing a knee-length tight skirt with a bright top and heels. There was a knock at the door. A guy who was about twenty-five with shaggy long blond hair stood there holding a notepad under his arm. He said he was an agent for some modeling company and was running late for his next gig, so he wanted to come in and use the phone. I told him no, but I was willing to make the call for him. He left, or so I thought.

If I Told You My Story

I was in the back bedroom changing my clothes when I heard footsteps—footsteps that would haunt me for years. This man had his way with me. I stood there sobbing and shaking. He threatened to kill me if I made a noise. He told me that he had never done this before and he was going to leave. After he left, I ran and locked all the doors and got in the shower and cried, letting the water run down over me to clean me, but I just felt I couldn't ever get clean. I got in my car and raced over to my friend Debby's house. She was furious. She phoned the police. After questioning me about my circumstances, they actually told Debby it was probably my fault. I wouldn't talk about this incident again for years.

I'M CLEAN

I did tell Cary. I don't know why. I wanted him to want me back and love me again. I still clung to a hope with him and allowed him to talk me into his bed once again after this happened. Why would he do that after what just happened to me? I wanted to be loved so badly. About a week later, Cary phoned me and asked if I had gone to a Doctor since the incident. I hadn't. I was too embarrassed too ashamed. He said he had a disease and thought maybe he got it from me. Scared, I went to the Doctor. It was a female Doctor. Through tears I explained what happened and why I was there. She examined me and said I was clean. I asked if she could write a note saying so that I could give to Cary. She was very angry with Cary's actions and was happy to write him a note.

This was the beginning of a change for me. I was ready to get outta dodge. I felt so alone. What was here for me? My family had their own happy lives. No one was ever going to want me with two babies from daddies who didn't support their babies. I was a burden to all, I felt.

An old high school friend came in town. Celia was married now to Vernon, a Japanese guy. Vernon had long black hair he wore back in a ponytail. They met and married in her hometown, Arkansas. Celia had a little girl, Angela, from a previous marriage, and now she and Vernon were going to have a child together. They said they were getting ready to move to Arizona, and I should join them. Tony was getting married in a month, and there was nothing left here for me. I had some friends at church and loved my church, but I felt I needed to start over where no one knew me. I only felt close to Jesus, but he seemed so far away right now, too. I read my Bible every day for comfort and strength. Celia and Vernon weren't into Jesus, but they loved me and I was welcome to join them in this adventure, so I said yes!

Cary wasn't happy about this at all. I was always his girl, and maybe someday when he was finished playing he would want me back. Now I was moving far away across the map and taking his baby girl with me. We said our goodbyes, and I sold everything I had. I rented a U-Haul trailer for the little bit I did want to take with me. My mom was terribly upset I was leaving and begged me to stay. My whole family just knew I would be running back in tears and begging for their help after a month. I saved up a lot of money to get me through my trip and now was ready to go. we stopped off at grandma Ginny's, (Cary's Mom) to say our good-byes.

Cary was Ginny's joy in life and Cary could do no wrong in her eyes, so she couldn't see why I would leave and take his baby girl away. Hello! This man, your son, wouldn't marry me or help provide financially for his baby girl. I want as far away from this man as I can get.

May 7, 1977: what an adventure this was going to be.

Again, all my life, even when I would step away from God, he still sent angels out to protect me. Now my babies and I were in a car with a U-Haul trailer, driving across the country. Bryan was five, and Sherah was a year old.

ARIZONA

We took our time and made it fun. Crossing the country from Midwest to the wild wild west! The transition of the scenery and the people was so different from what I was used to and grew up with. The closer we got into Arizona was beautiful. The kids loved it, and they were amazing little travelers! With this long travel it was like a vacation I'd never been on. Stopping at hotels, eating out. This was so much fun. It took us three days. I was so excited to make my new start in Arizona. Arizona was like a different country compared to where I came from.

I was supposed to meet up with a girlfriend, Amy, from St Charles that moved to Tucson

If I Told You My Story

before meeting up with Celia and Vernon. We were going to stay with Amy until Celia and Vernon arrived. Amy decided to leave Arizona and didn't tell me, so now I'm here a week early with no place to go. Money was running out and fear set in. Back home at the Sheep Shed, when out of town visitors stopped in needing a temporary place to stay, the Pastor made and announcement and someone always took them in. So I contacted the first Bible church I found in the yellow pages. I met with one of the Pastors at the church and spoke with him explaining my desperate situation. The best he could do was give me a phone number to the Red Cross. I left that church with tears and in disbelief. This Pastor was going to send this young girl that just pulled in from Missouri with a trailer hooked to the back of her 1967 Chevy Impala, with two babies out to the streets with a phone number to the Red Cross. I had to fight back the tears so my babies didn't know we were in trouble. I felt like God deserted me.

I called the Red Cross and they referred me to a place for Abused Women. They may be able to help me out.

I called them and they were willing to help me! You are only allowed to stay there a week and they didn't charge anything. During that week they would care for your children, but you must go search for a job. I never left my babies with them. Food and shelter was provided for you but you had to do your part of cleaning. We were given our own private bedroom. It had one bed and we all slept together. I wouldn't want my babies anywhere else but right next to me anyway. There were all kinds of troubled women there. Some with babies some alone that had just been abused by a boyfriend, husband or Father.

I phoned Celia in tears. She said we are going to leave now. It took them about three days to get there. Just in time when three really strange girls traveling together moved in.

SIN CITY

Vernon was not allowed of course on the premises, so Celia came in to get me. Celia had always been a great friend. She was always there and always cared for others. We all got along great! We found an apartment in Tempe, in "Sin City." That's what it was called where we set up our first home, we found out later. Life was changing for the better now. We had a two-bedroom apartment. Bedrooms upstairs. Vernon got a job right away and It didn't take me long to get a Dental Assistant job working at Dr. Pease dental clinic. Celia would stay home with all the kids. She loved children. I hadn't been in search for a church yet. After my experience, I was in no hurry.

Our apartment was situated with a pool in the middle of the complex and apartments in a circle around the pool. I can't imagine that being legal today. Celia was a great swimmer, so she taught all the kids how to swim. She was so patient and trustworthy. Weekends off were great to hang out at the pool. Here I was, 22-year-old cute little body in my bikini in the pool on the raft, getting as dark as I could. Luckily, I stopped soaking up the sun before it caused a lot of sun damage to my skin. We got to know a few of the neighbors, especially those who had children.

Sherah was turning two on July 11th. We had her birthday party at the Phoenix Zoo. A birthday party at the Phoenix Zoo in July is not exactly fun. It must have been at least 105 degrees.

If I Told You My Story

Directly across across from our apartment there lived about four very good looking guys sharing an apartment. After a few months the guys started talking to me. The one guy, Mark in particular latched on to me. Mark was tall dark hair and brown eyes and a competitive body builder. He started befriending my kids and buying them ice cream and candy. Mark said he had been watching me ever since I moved in and wanted to ask me out. He told me he was a private detective. Wow! That seemed so exciting and interesting.

Soon Mark asked me to a high school reunion party for Friday night. He seemed to be a nice enough guy so I accepted. Celia and Vernon were happy I was getting out and gladly offered to babysit. I got dinner for my kids and got them settled in for bed and met Mark outside about 8pm.

Mark had a red convertible. He had the top down. I got in and we started out. Seat belts were not mandatory therefore most people never wore them. I didn't have my seatbelt on, Mark noticed and insisted I put it on. His approach was somewhat aggressive so I put it on. I noticed Mark always checking out his rearview mirror. I thought it probably had to do with being a private detective, always on the look out.

The party was not a lot a fun for me being I didn't know anyone there. It was about 1am and I asked Mark if he could take me home. He told his friends he would be back he was just going to take me home. We pull into the guest parking of our apartments. Mark noticed a guy with long hair walking outside our apartment and wondered what he was up to. I just shook it off as, hey, we live in a college town in Sin City, it's a weekend, and it's probably someone walking back home from a party.

Mark walks me to my door and gets back into his red convertible. I realized I didn't have my house key but I would just knock on the door to wake Celia up to let me in. All of a sudden from around the corner a long haired guy approached me with a rifle! I screamed! He turned away from me and joined about six guys that pounced on top of Mark in his car. Mark had no chance. I'm pounding on the door. Celia and Vernon are not answering. I felt I was free so I ran next door to the neighbor we were friends with that had a daughter the same age as Bryan and Angela. I'm screaming and pounding on their door. They come and let me in. Now we see the police squad cars all over the property and in Mark's apartment. What the heck is going on. I phone Celia and she meets me at the door. We all go upstairs and watch the scene from my bedroom window.

What is happening?! The next day, Mark's brother came over and told me the story.

A BRUTAL MURDER

It had been about a year ago. Mark and Bobby were hired as "hit men" for a big time drug dealer out of Mexico. So that's what he meant by saying he was a "private detective." Mark and Bobby were friends and hung out with the same group of friends. Bobby was tall shaggy blonde hair and blue eyes. I'd never met Bobby, but heard about him and read the police report later. Mark and Bobby were very competitive body builders. Mark was apparently jealous of Bobby because Bobby got more attention from the girls.

They would sometimes meet clients at the Phoenix airport and make an exchange of brief case of drugs and cash at the terminal.

The drug dealer boss thought Bobby was taking some of the money before turning it over to him. So he hired Mark to check Bobby out and get rid of him.

They were all partying at Minder Binders. A sports bar in Tempe. In 1977, Minder Binders sat on the corner of an intersection all by itself. It was all desert behind Minder Binders. Mark made sure Bobby was intoxicated and gave him drugs. Not sure what it was he gave him, but Bobby was loaded and pretty incoherent.

Mark and another buddy took Bobby outside to Mark's truck. They were all sitting in the cab. Mark started questioning Bobby and shoving him. Bobby, not being able to sit up very well, was dragged out of the truck. Mark took a hunting knife and started yelling at Bobby and stabbing him. Pretty soon, Bobby's body lay on the ground in a pool of blood. He's now dead. Mark told his friend to help him put Bobby's body in the back of the shell

A Brutal Murder

camper of his truck. Mark and the other guy drove up to the White Mountains to dump Bobby's body. That was Mark's second mistake: bringing a witness. The first mistake was murder.

It was a year or so later when someone found Bobby's skeletal body with his boots and clothing still on him. The buddy that was with Mark got scared and witnessed the murder to the police. The police investigated, and found Bobby's body was where the guy told him it would be. The body was identified by dental records. The SWAT cops were watching and waiting for the right time to grab Mark. They knew I was an innocent neighbor girl, so that's why they never bothered me.

Mark's brother, Jessie was over wanting to talk to me about this murder almost daily. Finally Jessie asked me to lunch to talk about his plan. He wanted me to go out with the guy that squealed on Mark so Jessie and and his gang would follow us and knock this guy off.

This was conspiracy to murder! I didn't want anything to do with these guys ever again. The lease for this apartment would soon be up and we would move.

ARIZONA DUST STORM

One night I was up until past midnight. Everyone was sleeping. A big dust storm was just beginning to kick up. I thought I'd better put my windows in my car up. I was still new here and not used to keeping my car locked and windows rolled up. I say rolled up, because I didn't have a car with electric-powered windows yet so I had to manually roll up my car windows. We had unassigned covered parking. The parking was just right outside the kitchen door. There was a wall between the kitchen and the entrance into the apartment. On the kitchen side was a window that looked out to the parking lot. I opened the door, and closed it behind me just as the dust and tumbleweeds were really starting to blow. My car happened to be parked almost right in front of my door.

I started walking out to my car and saw a guy with long, sandy blond hair walking towards me. He was a few doors down from me. Fear struck me. I looked at him, but we didn't have eye contact because he kept his head down. I didn't have a good feeling about this guy. I got in my car and quickly put the windows up. As I was approaching my kitchen door to my apartment, his pace started picking up. I ran into my apartment and locked the door! With my heart pounding, I was leaning against the door. Even though I couldn't see in the kitchen, I felt he was peering into the window of the kitchen. The window was closed and latched. Something told me to make sure the arcadia door in the living room was locked.

The arcadia door in the living room was across the room. The arcadia door led out to the swimming pool area. I quickly walked over to the door and pulled the curtain back, and sure enough the door was not locked, and this guy was standing there with his hand on the door! Normally when I get scared, I freeze, but this time I forced a chilling scream! Luckily the children didn't wake, but Celia came running down stairs. I told her what happened. We didn't see anyone through the windows, but we phoned the police anyway. They were there in a flash. I suppose they were always hanging around Sin City. Me, Celia, and the policeman were all standing in the doorway with the dust storm picking up. Tumbleweed being tossed across the parking lot and sprinkles of rain beginning to fall. I began describing the guy and telling the police officer what happened. All of a sudden, a tumbleweed blew by and scared the daylights out of the three of us!

Our six-month lease was up, and we were ready to move out of Sin City.

We all moved to Mesa together. Both me and Vernon's jobs were closer there. It was barely six months when Celia and Vernon decided Arizona wasn't for them, so they packed up and moved back to Arkansas. Even though this Wild West was not working out so well for me either, still I wanted to stay and give it another chance.

ALONE AGAIN

Paying for everything myself now was difficult on the little income I made. Finally, I couldn't afford the electric bill, so our electricity was turned off. I had to get help. I was told to go to the welfare office in the town of Mesa to see if they could help me get my electricity turned back on. I sat in that welfare room for hours, it seemed. I felt very uncomfortable. I'm used to working for my needs, I'm not good with handouts. I would make myself work harder so I would not have to do this again.

Finally my name was called. They did indeed help get my electricity electricity turned turned back on. They gave me a couple months only then I would have to start paying for it myself. I was very grateful. I knew I had to move again to something more affordable.

I found an ad in the paper for a trailer for rent. It was $125 a month. It was a two bedroom tiny mobile trailer. This lady had three trailers on a lot that sat back off of Spencer street in Tempe. It was all desert around the trailers. I rented the one on the end. There was one young couple with a little girl about Bryan's age on the other end. A Mexican lady with about four children in the middle. This Mexican lady, Marcella, was so sweet. Marcella and her kids were between 4 and 12 years old. Her daughter, who was about ten years old, was mentally challenged. Marcella didn't work and had a married boyfriend. Two of the younger children were from him. He was a handsome man, but a dirt ball. I didn't like him. He was at Marcella's trailer quite a bit during the day. They would lock the children outside while they were inside having crazy wild sex. That trailer would be bouncing!

Alone Again

I got to be good friends with the couple that had the trailer on the other side, Mimi and Kurt and their little girl, Cara. Mimi had two brothers in Pine Top. They were contractors and built cabins in the mountains: tall, strong, blonde, blue-eyed guys. Bill was the oldest. I had my eye on him. But the younger, Roger had his eye on me and spoke up before Bill, so that left Bill out not wanting to interfere with his younger brother's interest.

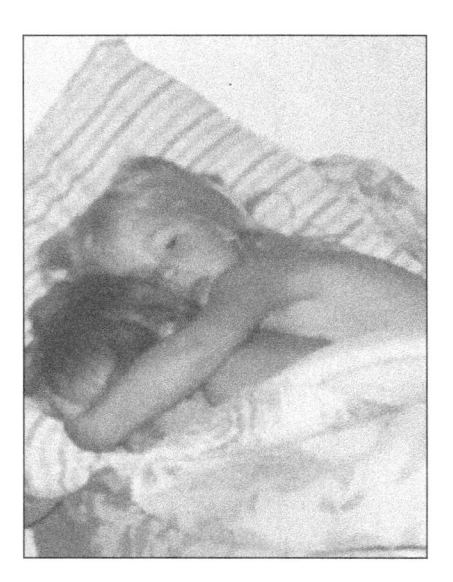

Me and my kids would go up with Mimi and Kurt to Pine Top quite a bit to visit the guys. Bryan and Sherah loved it there. Romance never blossomed between me and Roger. I just had no spark. I wanted to date Bill, but that just wasn't going to happen now.

A NEW JOB

I didn't like working at the Dental Clinic so I found a job working for another Dentist. He had a couple offices. One in Tempe and one in Phoenix. I worked both when I first started out then I transferred over to working the Tempe office. I liked working there. There were only three of us girls there. Joey ran the front office, me and Becky worked chair side assistant. Joey's husband was a policeman so when he was working she and I would take our kids to Payson to go sledding down the hillside. Packing hot chili and hot chocolate for lunch.

Becky's good friend was Andy Bombeck. Yep, Erma Bombeck's son. He was so cute!

If I Told You My Story

Funny and handsome. Long brownish hair and green eyes. Andy would come in the office on slow days a lot and visit with Becky. Andy told Becky he really liked me and wanted her to ask if I would be interested in going out with him. Me? I'm a nobody. I lived in an old, ratty trailer where the window on the door would fall out if you closed it too hard. And I had two children! Why would he be interested in *me?!* Nonetheless, he was, and after turning him down several times, I finally accepted. Andy took me to dinner and a concert. Rod Stewart. I was excited! I loved Rod Stewart!

Joey babysat while I went out on my first date. The concert was at the Amphitheater downtown Phoenix. A revolving stage. It was really fun. After our date Andy brought me home. I was terribly embarrassed. He gave me a sweet kiss good night and when he went to close the door as he was leaving, I was holding my breath that the window wouldn't fall out. It did. Andy is fumbling with the

A New Job

window and trying to get it put back in. I'm horrified. I'm reaching for my grey masking tape to tape the window back up and apologizing as I'm taping. Andy kept calling me but I wouldn't answer. I was not good enough to date such a fine, up-scale guy, especially if he was from a famous family. Finally, he gave up asking Becky.

RICK AND NICKY

Bryan got to be good friends with a little boy down the street who was in his class at school, Jacob. Jacob's Mother was a single girl too. Nicky was her name. Nicky was a school teacher. Skinny girl with long blonde hair and big blue eyes, a pretty girl. She was about six years older than me. Nicky was so smart, I wished I could be like Nicky. We got to be great friends. Rick was living with Nicky. Rick had curly long brown hair and brown eyes. He rode a motorcycle to work as a mechanic at a garage.

Nicky seeing I needed to get out of this trailer situation wanted to help me find another place. She lived down the street in one of the Six Plex apartments. It was so cute! Everyone that lived there was like family. Finally an apartment opened up. Nicky spoke with the owner and he let me rent and we moved in. Now my little family was part of this family. I loved it here! We all helped each other out with babysitting and having an occasional dinner together, or just chic chat. We were truly family. We shared holidays, had BBQs in the backyard. I was so happy!

I felt like this was the first place I could really make home for my babies. Bryan and Jacob would take turns spending the night with each other. They got to be really good buddies. Every time Bryan stayed at Jacob's, in the middle of the night there would be this little knock on my door. It was my little Bryan in the dark looking all around waiting for mama to let him in. Bryan just liked being home.

Even the trailer we lived in, I kept it spotless. I loved my homes to be clean and smell clean. I started buying furniture on credit. I usually found specials during holidays where they gave you time to pay it off without interest. I always took advantage of that. I made us a cute little home. I always loved the American Indian culture. My great great grandmother, Polly Snow was 100% American Indian. I loved American Indian art so I started a collection of the art I could afford. I would get dinner for the kids and clean up the kitchen then sneak out and go run the rail road tracks to decompress.

HOME

Our first Christmas at this new home was hard. I always got us a real Christmas tree. I loved the smell of the pine. I didn't have a lot of money. But I got enough money together to get the kids new pajamas and a couple new outfits. I bought a used train track for Bryan. Christmas Eve I would set it around the Christmas tree like I'd seen in movies. For Sherah I hand made her a rag doll. I named her Alice. Alice had red yarn braided pig tails and a dress. She had button eyes and a painted on smile. I worked long and hard at making that doll for my baby. I don't think my kids even knew we were as poor as we were. Christmas morning was still fun for them. You can't wrap happy and put it under the Christmas tree, but you sure can show happy through love.

I felt like the worst loser and now I've brought these two precious babies into my world of struggle. Hugging and kissing their little faces, I *always* told them, "it's gonna get better." They heard that from their mama all their life growing up.

I had come out of the hippie movement now, trying to work towards being more of the professional world. This new group of friends was still in the hippie movement. They introduced me to all their friends. There was one guy, Don, he was much older with a long Willie Nelson type look with the long blonde braided pony tail. Don had a large farm in east Mesa. We shared holidays together there. We would sit around the campfire and sing. Those that played guitar played and sang. Nicky played and sang. She was such a beautiful smart girl. We were such great friends. She always told me how beautiful I was and my prince charming would be coming.

Soon a friend of Rick's popped up, Don Barker. Don was very hippie. A very gentle spirited man. Don was early to mid thirties. Thin with long blonde hair and blue eyes. Don was very much into nature and eating raw or organic foods. Don was an Artist. Only because Don was single and part of our family and such a nice guy, I thought I should try to hook up with him. It just didn't work though. We couldn't get past friendship.

Don sort of became a squatter living in this huge house in Jerome.

Jerome sits above what was the largest copper mine in Arizona and produced an astonishing 3 million pounds of copper per month in the late 1800s. Men and women from all over the world made their way to Arizona to find work and maybe a new way of life. Jerome is a town in the Black Hills of Yavapai County between Prescott and Flagstaff. Jerome was founded in the late 19th century on Cleopatra Hill overlooking the Verde Valley, it is more than 5,000 feet above sea level. Jerome was once known as a copper mining camp. WW2 brought a huge demand for copper, but after the war, demand slowed. Jerome was dependent on the copper market bringing a closure to the mine in 1953 and Jerome became a ghost town. Jerome became known as the largest ghost town in America.

Today Jerome is a thriving tourist and artist community with a population of about 450.

At the time we were there, Jerome was a ghost town. There were a couple people living there in the old deserted homes. We stayed with Don in this deserted huge home that sat on the side of the mountain overlooking Jerome. At night there would only be the moon and stars for light. It was magical.

One afternoon I came home from work and there were all kinds of people at Nicky's apartment. Rick had been in a motorcycle accident. Someone ran a red light and Rick flew in the air. Arizona doesn't have a helmet law and Rick was not wearing a helmet. Rick died the next day. This is when we found out Rick was married. Nicky was not allowed at the funeral. Rick's Canadian wife came in and didn't want Nicky anywhere near her husband's funeral.

Nicky was crushed. Don was there to console her. Soon Don and Nicky became an item and they married. Don and Nicky had a baby girl and named her Summer. I was so happy for them, they were more suited for each other than I would have been.

CLOWN SCHOOL

I learned to do whatever I could do to make extra money for my little family. When Sherah was in kindergarten I met this guy, Arnold, he worked as a professional clown. Arnold was known in the valley for his clowning. He asked me if I wanted to earn some money being a clown. He helped me come up with a design. I wore a rainbow colored afro wig. Big floppy shoes and a big clown suit with polka dots.

I painted my whole face white with a rainbow across my nose over my cheeks, and big red lips. I worked at Motorola picnics at the park. Making balloon animals and painting designs on the faces of the children.

I was paid $150 for three hours of clowning. I started clowning for parties everywhere.

I taught Bryan how to make balloon animals and he became my helper. Once Bryan and I clowned for Sherah's kindergarten class at school. It was put in the Arizona Republic. The article showed a picture of me and Bryan and said I graduated from "Clown School!"

I soon got a better paying dental job in Scottsdale working for a group of Oral Surgeons. There were three Oral Surgeons.

This is where I would meet my life long sister friend, Nancy. Nancy was in charge of the office. We quickly became best of friends. Nancy was a Christian and invited me to go to church with her. I hadn't gone back to church in a while and really missed it. I started taking my children to church then. In the summer I sent my kids to camp with the church. They always had a good time.

I loved this new Christian friendship. I met a lot of friends through Nancy. We all would drive up to the White Mountains to ski in the winter. Go water skiing in the summer. And just hang out.

Soon Nancy was going to move on and out of Dentistry. Nancy felt a calling from God to move to Africa as a missionary. It was hard to see Nancy move away. I quit going to church again. One of the doctors was going through a divorce. Pretty soon he and I started secretly dating. That secret didn't last long and it was best I leave the office.

I got a job teaching dental assisting at an Arizona College of Dental Assisting. I loved dentistry and I loved this job. I loved teaching and sharing all the different fields of Dentistry.

Another Dentist used to come into the Oral Surgeon's office to observe oral surgery before I had left. He, Dr Bishop was just finishing dental school and thought he may want to go into Oral Surgery.

Summer time my Mother and Hank, would drive out to Arizona and get the kids and take them back to Missouri for a few weeks. I worked a lot while they were gone. This is when the romance started with George. After George and I started dating, Dr. Bishop and his wife, Marla, would become our good friends. We would tailgate at football games and other sporting events. We got together for dinners and holidays. They would eventually become my best friends, even to this day.

George was one of the best men I'd ever dated. I didn't want to date him, but he kept chasing me and begging me to go out. Again, I thought, "I'm a country girl with two kids with nothing to offer, so why me?" George reminded me of the Prince of Wales, who left the royal kingdom to marry his common love.

My first fancy dinner with George was a beautiful little Italian place in Scottsdale. I didn't drink much alcohol, so it didn't take much before I felt the buzz. We were sitting at a little table with the white linen table cloth and a candle in the middle. After a few sips of wine, I reached up to take a sip of water and knocked over the water glass. I was terribly embarrassed and wanted to leave. George wasn't embarrassed at all and just asked the waiter to get us a new table.

George taught me so much and encouraged me to further my education. I finished my education and started earning credits toward nursing school. It was hard to go to school with the two kids. They were in primary school and I couldn't leave them alone for long periods of a time, so I ended up putting school on hold.

AWA

Once when we were out with Dr. Bishop and his wife, Marla, they asked me if I would be interested in being a flight attendant. "Yes! I'd love to." I used to think about it when I was a young girl lying in fields of grass. I would look up and see airplanes fly over and wonder what it would be like to be in one.

A new airline was just starting to hire flight attendants. Marla worked for America West Airlines in the office. She got me an interview with the hiring department who happened to be her friend, and I was hired.

I started training for AWA January 1984. We were not paid during our four or six weeks of training. Also, we were required to purchase $2000 of AWA stock. They gave us the option to take it out of our paychecks until it was paid off. We were trained to be cross utilized. We were called CSR's. Customer Service Representatives. We were cross-utilized, which meant we were trained in all facets of customer care. We would be working ticket counter, gates, baggage claim, baggage handling, ramp, catering, and flight operations as well as being a flight attendant. Every day or two was something different. We did ticketing on board. At this time, we only had Boeing 737 airplanes. Nonetheless, training was intense.

During training we were weighed in weekly and tested at the end of every week. If you failed more than one on the exam, you were out! We watched people dropping out one by one. Some just because it was too intense and they didn't want to do it, and others because they failed the exams. When I came online, we only had six airplanes.

Other than the little pay, this was a great job—the best job I'd ever had. I liked flying the best. It was pretty crazy though. A passenger could decide to fly to Los Angeles at the last minute, so he could hop on the airplane and pay once he got inflight. The gate agent would make their reservation right there and put the passenger on. Then we would come down the aisle with a cart collecting ticket monies in the form of a credit card or cash. You paid on board. There would be three flight attendants. One collecting the ticket monies, the other two serving drinks. When you're flying a short hop, this got to be

If I Told You My Story

challenging. It all had to be accountable per passenger by the time we landed so we could give the monies to the gate agent before we either took off for another flight or left for the hotel.

I would make my lifelong friends at America West Airlines. I flew for free as well as my children and parents. Then I was given passes to give to family and friends. They paid only $20 a pass round trip. It was all standby flying.

AWA started growing so fast that they realized they had to do away with being cross-utilized. We now got to choose where we wanted to work. I chose flying, and that's when I became a full time flight attendant.

I would be out of town about two nights a week. George would babysit the kids for me. Sherah was in first grade when I started working at AWA, and Bryan was in fifth grade.

Our fleet was growing so fast at AWA. We bought 757 aircraft and took them to the Big

Apple. New York City! What an exciting city this was. It was almost like my second home, I was there so much. Then we bought the Boeing 747s and started service to Honolulu and Japan.

George became insanely jealous, thinking I was meeting other men. But still he wouldn't marry me. Only when I told him I was leaving him did he say, "No, we'll get married." But then he would never propose. Finally, I had enough. I was feeling cheated in life. I wanted to be married. What was wrong with me? Why didn't anyone want to marry me? I was thirty-four and told George I was leaving. We broke up for good this time. When George and I broke up, Bryan was almost 16. He was beginning to give me problems that were difficult for a single mom. I gave him choices. Live by my rules, or live with your Grandma. He chose to move to St. Louis to live with my Mother and Hank. I hated having my boy gone. I cried myself to sleep. Of course

Bryan straightened up and finished school. Eventually Bryan would make his way back west and end up in California.

I worked a lot and had to be out of town quite a bit. Sherah was twelve. We lived in a condo. There were college guys who lived upstairs and helped me out a lot, like little brothers to me. And a married couple, Jeff and Christie, who lived next door. Between all my neighbors, they looked after Sherah when I was out of town. I made meals for Sherah to eat while I was gone. Sherah was the perfect child. She never gave me trouble and was a great student, making almost all A's. She was a cheerleader all through school, as well as through college.

DESERT STORM

In the last months of 1990, the United States participated in the defense of Saudi Arabia in a deployment known as Operation Desert Shield. Over 500,000 American troops were placed in Saudi Arabia in case of an Iraqi attack on the Saudis. Traditionally, Iraq was an ally of the Soviet Union, who held a veto power over any potential UN military action. Looking westward for support for their dramatic internal changes, the USSR did not block the American plan. The UN condemned Iraq and helped form a coalition to fight Saddam militarily.

Operation Desert Shield which turned Operation Desert Storm, broke out and the government commandeered our 747 planes to take military to the Gulf. AWA announced we would be taking troops to the Gulf. Flight attendants were chosen by seniority to work the flights. If you wanted to volunteer, you signed up. The pay was great, so I signed up. I wasn't happy about having to leave Sherah again, but I always wanted to provide for her for all she wanted so she didn't feel different from her friends that had two parents.

It crushed me when Sherah would come home in tears because she didn't have what her friends had. She never complained, but a girl is a girl. I used to tell her, "Sweetie, you just take care of your God-given beauty and make sure you take the best of care for your hair and your body, and you will always be beautiful!" "So what if you wore the same jeans and jean shorts, I'll make sure you always have new tops to change it up; you do your part and I'll do mine." Sherah had beautiful long blonde hair and big brown eyes, a beautiful girl with lots of friends. Always sweet as pie!

Sherah was only fifteen when I volunteered to go to Saudi Arabia. I had people taking care of her. Primarily Jeff and Christie was who I relied on the most. A strong Christian couple that loved me and Sherah like family. Then she would stay with her best friend's family, Christine, while I was gone too. I would be gone at least a week at a time.

When we first started flying the troops, the flight attendants weren't replaced in Brussels like the pilots were. We would pick up military servicemen and women up all over the country and fly them to Saudi, Riyadh, Kuwait, or Jordan, stopping in Brussels to refuel and replace pilots. When we stopped in Brussels, no one was allowed to leave the aircraft, especially the servicemen and women. We were there only long enough to be catered and re-fueled. Some of these guys didn't have time to phone their wives or families before we left the base in the USA. They weren't told, apparently, where they were going. So they would write a message on a piece of paper, and I would hand it to the gate agent, and they phoned the families to read the note to them. These messages brought tears to my eyes. Words of love, and "I miss you already."

It was always interesting to listen to these servicemen and women of all different walks of life. They started giving us their medals and pinning them to our uniforms. I have tons of medals, and I have a green beret hat, plus dog tags and photos. I even ended up with the American flag they flew on base in Saudi Arabia. It was crazy seeing rifles and bazookas on the airplane. These guys called us "Flying Angels." They wrote poems for us and wrote us letters and sent photos of what was going on in Saudi Arabia.

We were doing Saudi turns. Taking troops over with a stop over in Brussels and ferry the plane back to Phoenix, AZ. To ferry is to have no passengers on board. We were on that airplane for four to five days at a time without a shower!

On the ferry flight home we got out of uniform and into our comfy clothes and would jog around the airplane for exercise. The girls would help wash each other's hair. Having clean hair just made you feel clean all over. We had buckets of water and a pitcher of water to pour over our hair. I massaged feet. We told stories and slept. When we stopped over in Brussels, we always got off and bought some of their famous Belgian chocolate my daughter loved. Soon the FAA discovered the flight attendants were working these flights without a layover, so that changed everything. Now we would layover and do a crew change in Brussels like the pilots did.

Driving through Brussels in the middle of the night on a crew bus was interesting. The bus driver took us through the famous red light district. Amsterdam's red light district (aka De Wallen) has been a familiar haunt for pleasure-seekers since the 14th century. Though certainly not an area for everyone, the red light district has more to offer than just sex and liquor. Underneath its promiscuous façade, the area contains some of Amsterdam's prettiest canals, excellent bars and restaurants, and shops of all kinds. It also consists of windows with sexy girls, dressed in eye-popping underwear.

When we landed on base in Saudi Arabia, a soldier always came on board and demonstrated how to put on the gas masks which they provided for each of us. He would tell the Captain what direction to take off should Saudi Arabia be attacked while we were there. We were allowed to get off the airplane for a short time before we took off. The base was hot and desolate. There were tanks all around. The soldiers loved showing us around base and allowed us to go down into the tank. Only one person could fit at a time. I could see how claustrophobia could set in really quickly.

We were there every week, so I asked the guys what they missed that I could bring back for them. No more cookies please. A young, 23-year-old guy piped up and said, "I'd like vodka." Saudi Arabia is a dry country, meaning alcohol was not allowed. Having a dental background, I prepared a dental package for these guys and put dental floss, toothbrushes, toothpaste, and a couple large bottles of Listerine in a box. I poured out the Listerine, rinsed the bottle, poured in the vodka, and added food coloring. Of course, I put cookies in the box as well. I thought, "This young guy could die tonight, and if this is his only wish, I'm going to get it for him." I really should have added a Bible, not vodka! Not so smart on my part. If I had gotten caught, I wouldn't be here to write about it I'm sure.

It was very early in the morning when we landed in Saudi one day. There was a lot of activity going on in the air as well as on ground. You could feel something was up. By the time I got back home, it was on the news. On February 24, the ground war began.

Although the bombing lasted for weeks, American ground troops declared Kuwait liberated just 100 hours after the ground attack was initiated. American foot soldiers moved through Kuwait and entered southern Iraq. Of course I was emotional and concerned about the safety of all these soldiers we'd gotten to be so close to.

The war didn't last long and soon we were bringing troops home. This was almost as emotional as it was taking them over. When we landed at base in the USA, a red carpet was always laid out and CNN or some news team would be there to film these service men and women getting off the plane. I did this for about six months: taking troops over and bringing them home.

This was a big adjustment for me and my body I can only imagine what it was like for them. The US Air Force General gave us medals with a ceremony when it was all over.

AFRICA

After the war, America West went into bankruptcy. What was I going to do if I lost my job? I didn't want to go back into dentistry.

Now Sherah was graduating from high school. Her graduation gift was a month in Europe with her best friend, Jane. Jane's parents and I would take turns having the girls phone home collect to see how they were doing. Cell phones weren't a common thing to have yet. Sherah was enrolled at ASU in the fall for journalism. I would really be an empty nester now and wasn't looking forward to it. Was I going to have a job tomorrow?

I decided to go to school for makeup artistry in Hollywood, CA. I was able to maneuver my schedule with vacation time and just taking time off and move to Orange County to go to school in Hollywood at the Bill Blasco School of Makeup Artistry. I moved in with my sister and drove with the California traffic everyday to Hollywood. About 2 hours each way. School was a lot of fun. I learned how to do bald capping, make a beard, do beauty as well as special effects make up. My forte was special effects makeup.

After finishing make up school, I worked in Phoenix doing make up jobs at whatever I could get and still flew as a flight attendant. Sometimes I would travel to do makeup. I sometimes flew into Burbank to visit the make up stores to get all my supplies such as a gallon of makeup blood and taking it back with me on the airplane. I could do that back then, as well as carry my big makeup case. It was a lot of fun and long hours working on film and then doing weddings for extra money. Makeup was exciting, but I didn't want to move to Los Angeles to really get into the entertaining field of makeup artistry. I did makeup for about ten years.

Sherah finished college at ASU and now looking for a job as a sports caster. She did little jobs around town but was ready to really go big. She was offered a job in San Bernardino, CA. She *just* moved there and I took off for Africa. I made several visits to Africa but one of my favorites was with the girls.

I had met a guy on my flight who was living in Africa: a tall, blond, blue-eyed man. Robert had a business that put up cable towers in remote areas of Africa. I was reading the *Left Behind* books at this time, and it sure made me think of the preparation for the end days as it's described in those books. I was starting to go back to church again. I'd always wanted to go to Africa, and now I was living that dream.

Africa

My friend Nancy was living there as a missionary, so this was a perfect time to go, and I could stay with Nancy and see Robert too. The first time I went there was with two other flight attendant friends so we all stayed with Nancy.

Africa was amazing!

One of the best places I'd ever experienced. I loved the people there. Before we did a ten-day picture safari, Robert and I got together and did some traveling. We went to a remote village and took a small row boat across the water to a little island. The waters were muddy; you would expect a hippo to come dump your little boat over any minute.

We went to another village. The natives had painted up faces and huge holes in their ears and rings in their nose. Photos were a taboo, unless of course you gave them money. The children from the village came down to drink from the muddy waters. I just wanted to take them all home with me. The people in Africa were so friendly. I could write a whole chapter on Africa alone.

One afternoon back at Robert's place I was waiting for him to come meet me before we went to meet my friends for dinner at the Carnivore Restaurant. He came racing in with his driver in the jeep. A gang was after them to kill Robert. I never really found out what that was all about, but he left Africa for good not long after that.

Back in town in Nairobi after meeting up with my friends, Donnah and Bobbi, we did

Africa

a safari. One night we stayed at the famous Tree House. We each had our own rooms. In the middle of the night, the watchman would buzz our rooms to let us know when the animals showed up. Below us was the watering hole, where the animals would come and bathe and drink. One night in the middle of the night, I got up to go to the restroom. We all shared restrooms. The ladies had theirs, and the gentleman theirs. I was wearing a long, white cotton gown. As I quietly went into the restroom, an English lady was getting ready

If I Told You My Story

to leave. The English have many folk ghost stories, and I'm sure when she saw me, whom she wasn't expecting, I looked like a ghost from one of her recent stories. I thought she was going to have a heart attack right then and there!

We moved on to staying at the tents. This campground had a wall around it to keep the predatory animals out. We three girls shared a tent. The tent was very comfortable and had a restroom with a toilet and shower in the tent. That's my kind of roughing it. We slept on cots. The campground would put large hot water bottles in our beds to keep them warm for us. When we first got there, the grounds lady told us that the monkeys would be coming out around 5pm and to be cautious and keep our tent zipped up, as they would steal items from the tents.

One evening at about 5pm, we were sitting in front of our tent, having tea. Pretty soon, we heard all the monkeys chattering,

Africa

swinging from branch to branch, and sliding down tents. One monkey sat on the bar atop our tent. He just sat there looking back and forth at us girls. Giggling, we thought he was adorable. This little guy squatted right above me and was just staring. I looked up at this little excited monkey. Yikes! Is this what the lady was trying to warn us about? We were cracking up! I think he got embarrassed because he ran off. Ha! I can't get a man to like me but I was sure I could get a little monkey! What an exotic trip Africa was with a lifetime of memories.

One morning we scheduled a private breakfast safari. Our jeep broke down and just a few yards away was a mama leopard with her babies shading under a tree. Ive never seen a flat tire be changed so quick as our driver changed ours. Leaving Africa we brought home a lifetime of memories.

Robert and I continued our relationship through phone calls and emails, and sometimes I would meet him wherever he was going to be in the world, in London or wherever. We planned to meet in Seychelles and marry. It may be a year from now, he said. Robert's contact with me began getting less and less frequent. What was up? "Just tell me, and I will go away," I said. The questions with his silence were what killed me. I found out Robert was married! What?! I never wanted to see his face again.

It was now Y2K. The Y2K Bug was expected to completely shut down computers around the world. While many were ready to party like it was 1999, many others predicted catastrophe at the end of the year from a small assumption made long ago when computers were first being programmed. The Y2K (Year 2000) problem existed because most dates in computers were programmed to automatically assume the date began with "19" as in "1977" and "1988." But when the date was to turn from December 31, 1999 to January

1, 2000, it was prophesied that computers would be so confused that they would shut down completely. I came home from a late night trip, and there just so happened to be a blackout in my neighborhood. Of course, I was thinking it was the Y2K shutting the city down. It was a little frightening going into my condo, but it was just a coincidental outage.

A SCARY VISITOR

Several months went by, and I started thinking about Robert. He disgusted me; get out of my head! Why was I thinking about him? I went to work and was flying with a friend, Kim. It was jump seat therapy time, so I began telling Kim my story about Robert and how thoughts of him were dancing in my head twenty- four -seven. We call it jump seat therapy because flight attendants spill their stories to anyone sitting in the jump seat next to them.

Kim was in the jump seat with me, reading a "new age" book. She tells me, "Oh, you were supposed to fly with me today. I'm reading this book, and I'm on the chapter explaining

A Scary Visitor

what you're going through. Go home tonight, have a glass of wine, and tell Robert's spirit to leave you alone. It's his spirit visiting you." I had read enough of the Bible and gone to church enough to know I did not want to talk to a spirit and I wasn't going to have a glass of wine, but I did go home that night, and I did tell Robert to go away and leave me alone. I was not interested in him any more, so he needed to leave me alone. Well, I had just opened the door to the spirit world.

That night at 3 am, it started. I had diamond stud earrings that I wore all the time. I almost never took them off, not even to bed. I have an old antique mantle clock my mother had given me. She bought it for 50 cents at an old country auction when I was a little girl. I'd become accustomed to wait for the chime of my clock to tell me what time it was if I would wake in the middle of the night. It chimed once on the half hour and then every hour on the hour. I was woken with a burning sensation in my left ear. I start rubbing my ear and turning side to side. I heard a 3 o'clock chime. I got up and took my earrings out, thinking it was my earrings rubbing, and then I went back to sleep. The next morning, I was sitting on my bathroom vanity as I always did to put my makeup on.

My ears had been pierced since I was 13 years old. When I was probably 15, me and a friend got a second hole pierced in each of our ears, so now I had two holes in my left ear and one in my right ear.

I was looking in the mirror and seeing a third hole in my left ear! My ear was red from the burning sensation. I was no longer wearing earrings. I went to my bed to see if I would find a scorpion or blood, or anything that could have caused me to get a hole in my ear. There was nothing!

I went to work and asked the flight attendant I was flying with if she saw anything on my ear. She looked at the back of my ear

A Scary Visitor

and told me there was a little scab as if I had just pierced my ear. I forgot about it and went home. That night, same time, same place, I felt the burning in my same ear, only this time I wasn't wearing earrings! I started rubbing my ear and turning from side to side.

The next night I woke again at the same time, but this time because my body was paralyzed. I couldn't move. I felt something massive next to me. I could "feel" it, but not physically, and I couldn't see it because I couldn't open my eyes. I felt the heat of something next to me. It was huge! I couldn't open my eyes or move. I couldn't even wriggle a toe. From the time I was a very little girl, my mama always taught us kids, "If you're ever afraid, you just say the name 'Jesus,' and that fear *has* to leave." This is scriptural too, by the way.

There is no fear in love; but perfect love casts out fear, because fear involves torment...

1 John 4:18 NKJV

If I Told You My Story

So I'm trying to say, "Jesus, help!" I can't get any words out because my body is paralyzed. I hear voices speaking to me. Not audible voices. Voices in my head. It's like a good angel and bad angel sitting on my shoulders whispering in each ear. The bad angel is whispering in the most angelic voice, "Rhonda, it's okay. Go back to sleep." These truly are the voices I heard. My name and all. My eyes are heavy and hard to stay awake now, I'm drifting. The good angel says boldly, "Rhonda! Don't fall asleep! Keep praying!" Finally, I forced myself to keep praying and keep trying to force the words out of my mouth, "Jesus, help!"

Pretty soon I was getting some sound to come out of my mouth. Pretty soon, I was sputtering, "Jesus, help!" Pretty soon I was speaking loud and clear, and I yelled, "Jesus, help me!" That very second, my body was mobilized, and I kept praying. For three nights in a row this happened where my body was paralyzed.

A Scary Visitor

For the next three weeks, I would experience some strange happenings. I was talking to my best friend, Theresa, one morning about what happened, and as I was explaining it to her, our phones got disconnected. I rang her back and she said, "Rhonda, this is freaking me out." One afternoon I was on my computer, typing an email to Klove, a Christian radio station. I'm telling them my story and asking for prayer and asking what they thought. Halfway through my email, my computer became disconnected! It was as if it was unplugged from the wall. I could *feel* evil breathing down my back. I spoke out loudly, rebuking Satan.

One night I was awoken by a sound of scuffling outside on my balcony of my bedroom. Of course it's 3am. My condo was on the second floor, and it is a very quiet neighborhood. I never heard neighbors. I was in a deep sleep. Normally I would wake up scared to death. I'm a big scaredy cat. The thought came to mind of Daniel 10. Daniel was fervently praying for his people for three weeks for forgiveness of their sins and terror was upon them. Gabriel, an angel of God, came to Daniel and told Daniel, "Your prayers have been answered the moment you prayed and I have come to respond, but I was held up by the 'Prince of Persia,' who is Satan, for twenty-one days. Then Michael, one of the chief princes, came to help me."

In my thoughts, like a voice speaking to me, this noise on my balcony was as if it was Gabriel fighting evil off for me on my balcony. I fell back to a peaceful sleep and didn't hear anything again.

It had been about three weeks now. I was flying what we called a triangle turn. It would be something like, Phoenix, Denver, Las Vegas, Phoenix, and we would get home at something like midnight. Well, this particular night, there was weather delays. Now we weren't going to get home until 3 am! I was so scared to walk into my dark condo alone

A Scary Visitor

at 3 am! I was flying with a Christian friend, and he told me to go home and read Psalm 23. I grew up with the Bible, so I knew exactly where to find Psalm 23. Great advice!

I walked into my condo and flipped on every light. I turned on TBN, the Christian television station. Still in my uniform, I grabbed my Bible and opened it up to Psalm 23 and started reading. I read:

You who sit down in the high God's presence, spend the night in Shaddai's shadow and say this: "God, you're my refuge, I trust in you and I'm safe!"

That's right, he rescues you from hidden traps. shields you from deadly hazards. His huge outstretched arms protect you, under them you're perfectly safe; his arms fend off all harm. Fear nothing: not wild wolves in the night, not flying arrows in the day, not disease that prowls through the darkness, and not disaster that erupts at high noon. Even though others succumb all around and drop like flies right and left, no harm will even graze you. You'll stand untouched.

Yes, because God's your refuge, the high God is in your very own home, and evil can't get close to you. Harm can't get through the door. He ordered his angels to guard you wherever you go. If you stumble, they'll catch you, their job is to keep you from falling. You'll walk unharmed among lions and snakes.

If you'll hold on to me for dear life, says God, I'll get you out of any trouble. I'll give you the best of care. If you'll only get to know and trust me. Call me and I'll answer, be at your side in bad times; I'll rescue you, then throw you a party. I'll give you a long life, give you a long drink of salvation.

Psalms 91—MSG

Did you catch this? I wasn't reading Psalm 23 at all! I do believe this was God the Father

A Scary Visitor

standing before me and speaking these words to me. I kept reading this over and over again, my face wet with tears flowing. I rebuked Satan to get out of my house and leave me alone, in the name of Jesus! I didn't even realize I wasn't reading Psalm 23 until the next day! Then I knew it was the Lord Jesus Christ speaking his words of comfort and protection over me. This was the last time I would experience terror again.

FLYING PRIVATE

Some of my flight attendant friends started flying private on the side so I got pulled into that and flew private on the side. I was hired by Swift Aviation as a flight attendant to fly private jets worldwide, as well as domestic.

Net Jets leased some aircraft from Swift, so I was actually flying for Net Jets. This was exciting and I loved it. I really enjoyed pampering my clients. I would get my orders for my flights sometimes a day before and sometimes it wasn't until a few hours before. I ordered all the catering for however long I would be with my clients with no mistakes, and it all had to be ready way before the passenger stepped foot on the aircraft. Sometimes it was preparing breakfast and lunch or just dinner or just lunch or snacks and lunch. I would greet my clients and help them with their luggage on the aircraft and hang their coats. We would fly celebrities and anyone who could afford a private jet. I almost always was tipped very well as they departed.

Net Jets and Swift aviation split, and now Swift began selling the Embraer airplanes. This aircraft was made in South America. We used some of their pilots as well as our own. I flew with Captain Adam the most. I always felt comfortable when Adam flew, no matter where in the world we were flying. I trusted him that much.

I was senior enough at America West that I could cut my hours and do another job, so this allowed me to fly private. I flew all over the world. The Embraer could be sold as a commercial or private jet. My boss would take photos of me in the plane for advertisements. At 5' 3", I could stand up without my head hitting the ceiling. It made the aircraft look larger. This aircraft was comparable to the 4G Gulf Stream. It was a very comfortable airplane.

We were a great team. My boss always flew with us. I made sure the airplane was white glove ready. I put fresh flowers on the dining table, made sure snacks were available and appealing. Fresh fruit was always on the table. Even the toilet paper roll was folded neatly to a point.

There was a potential client in England who showed a great interest in buying one of our planes. He needed to get back to Greece so we flew him there. His yacht was in Rhodes. He invited us all to join him on his yacht for lunch. It was early morning when we left London. We had a beautiful lunch on this incredible yacht. Adam and I walked the island after lunch. It was the most beautiful place I'd ever seen, with the most crystal blue waters. We flew back to England by dinner time.

It was my first time flying into Saudi Arabia selling airplanes. I asked my boss if I was allowed to bring my Bible. He said no, it was forbidden. This really saddened me. I took my little brown Bible all over the world with me. My Bible is my comfort and strength when I am scared. I prayed and asked God to bring me at least one Christian believer in this Jesus forsaken part of the world. I would go to the Internet Cafe and type in bible.com and read God's word. If I would be caught, who knows what would have happened to me. I didn't want to be disrespectful to Saudi's laws and their religion, but God is bigger than they are and I knew God would protect me from harm and I needed strength from his word.

Women in Saudi Arabia must wear a burka if they were married, or single women must wear an abaya. A burka covers all of you so your eyes are the only part of you that is exposed. In an abaya, your whole face is exposed. My boss had ordered an abaya for me to be delivered to the airplane once we landed. My boss is English and lived in Saudi Arabia for years at one time, so he spoke Farsi fluently, and we were happy about that. When we landed about 10pm, something was not updated with my boss's passport. They took him into the FBO. We had to stay on the airplane. They kept Dave in there about four hours. We were getting concerned and glad he could speak Farsi. Finally he came out and all was fine.

My Abaya was brought to me and we took a taxi to the hotel. Women are not allowed to sit in the front seat only men. We usually stayed at the Hilton hotel. I could see from my hotel window the building about six feet away with bullet holes in the building and bombing debris. I took a breath, said a prayer, and went to bed.

Women weren't even allowed to walk the streets without a man with them. Women didn't usually come out until 4pm. The mall had a whole section for only a women's presence. The shopping was incredible. Women usually wore either nothing or sexy lingerie under their burka or abaya. It was usually too hot to wear clothing too underneath that black, hot abaya or burka.

Our first night out for dinner was at the home of a Saudi diplomat who was a friend of Dave's from when he lived in Saudi Arabia. It was a garden party, outside in the beautiful garden, and it was catered with the best of food and wine. There were a couple of older women there, the wives of the diplomats. We had very interesting conversations at the dinner table. We learned about current events that are never talked about on the news in the USA.

I mentioned my daughter was home alone and I missed her. The wife took me into her home to the phone and dialed up the phone number for me to talk to Sherah and then left me for a private conversation. We were going to be in Dubai the next day. The Prince of Dubai was interested in our airplane. The ladies said, "Oh, the prince is going to love you!" They said, "If you stayed here, you would be married tomorrow." I'll bet so, I thought.

Still in Saudi, me, my boss and the two pilots, Adam and the pilot from South America, all men except me, went out for dinner. We walked into a little restaurant and seated ourselves. It was obvious we were Americans. The waiter came to our table apologizing and saying that I couldn't stay there. I must go to the family section, which was outside and upstairs. I got kicked out! Oh! I could never live here. I've been way too independent my whole life and now I'm getting kicked out of a restaurant because I'm female?! We all went upstairs to the family section. The guys got a kick out of that.

Women aren't allowed to eat with the men unless they are in the family section, which is separate from the dining area.

The tables were all booths with curtains pulled around them. Out of respect for her husband, a married woman will not uncover her face. She'll lift up the flap on her burka to eat.

Saudi is an Arab nation, and a lot of foreign workers go there. It is not a free country. You cannot just go there and see the sights. You can only enter on a business visa. It is mostly desert and that does not mean sand as most people think. It is dried dirt so dust storms are common place as its dusty. It is, however, a safe place for the most part, as long as you do as they do and abide by the rules: no drink, no girls, no normal fun as you may know it. Shopping is the best there. There are no taxes and things like gold and silver are at cost electronics. Nowhere is cheaper, not even Hong Kong. The beaches are beautiful, but you can't see any women; they are hidden. They wear all black gowns and veils. For an outsider, it is a very strange place. Just about everything we do in the USA is illegal there.

I was left alone during the day when we weren't meeting a client to show the Embraer to, so I would walk the cobble streets of Saudi. I always prayed for the Lord's hand of protection to be over me. I was trying to cross a street and this old man with only one tooth in his mouth and a big grin on his face leaning out of his driver side window saying who knows what to me. I'm trying to ignore him and traffic is getting backed up, with horns honking. It's a one-way narrow cobblestone street and I'm terrified now. Trying to ignore the old man, I'm thinking GO mister, before you get us both killed! Finally he moved on. I walked among the little cobblestone streets with vendors on each side. I made a few purchases, of a couple silver bracelets and a few other things.

I did love it here. It was so old with the cobblestone streets I expected Father Abraham to walk around the corner any minute!

ANSWERED PRAYER

Saudi Arabia is a great place to get gold. I went walking to the market one day to the gold district. Streets filled with tents everywhere, selling gold and almost anything you'd want in jewelry. I stuck out like a sore thumb, being a single American girl. Everyone I met wanted to talk to me. I sure got a lot of marriage proposals that day! I did purchase a lot of gold bracelets, Arabic coffee sets, and other souvenirs. And while I was there chatting with one of the owners of a shop, he brought up Christianity! My prayer was being answered! Being in this part of the world where Jesus is not accepted, I felt so isolated and really learned to lean solely of God and his presence.

I wasn't given the opportunity to join in on a weekly Bible study or go to church to worship my King in Saudi. Saudi has only about 8% Christian. It was only me and God, and now God answered my prayer and brought a Christian to me so I could talk about Jesus! It certainly brought a tear to my eye. God is a detail God.

That night I was hungry and ordered room service. I turned the TV on. There was a kind of celebration going on the TV. When the young guy brought me my food, I asked him what the celebration was and what were they saying. It was a religious celebration and he wasn't sure what they were saying. They were speaking in an unknown tongue to him. He looked at me and confessed he was Christian. *What*?! He explained to me his Mother taught him about Jesus and and he loves and worships Jesus. He and his family go to an underground Christian church to have fellowship with Jesus. In Saudi Arabia, their

Sabbath is on Saturday, and their Sunday is like our Monday. It was hard for this guy to get off work on Sundays to go worship his God, Jesus Christ. My eyes filled with tears. We prayed and he asked if he could send me up a basket of fruit as a gift. I accepted.

As we were leaving the hotel, I went into one of the shops that sold jewelry to look around while waiting for my crew. I started a conversation with the young guy behind the jewelry counter. I don't recall what we were talking about, but I said something about his religion. He looked at me and said, "I serve the same God as you do." I asked him what God he thought I served. He said, "Jesus Christ." I was stunned!

My God not only brought me the *one* Christian I prayed and asked for, he brought me *three!* This restored my joy. God seems to always work in threes in my life. A detail God he is!

Answered Prayer

The airport at Saudi we flew into was a small airport. It was rare to find a woman there. My boss and the pilots went out to prepare the airplane, and I waited inside the terminal. I had to go through a security screening used only for women. I went through this little dark room. There was a woman sleeping in there. She quickly got up and patted me down before I walked through the screening. I was waiting in the waiting area for Adam to come get me. A young guy of about 18 years old saw me and came running up with excitement to talk to me. He was the catering guy. It was very rare to see a woman in the airport, let alone an American woman. He was so excited. As I was about to remove the airplane pendant on my uniform to give to him, I saw fear cover his face. There was a little pink room with a door across the way where women should wait behind a closed door. He told me I had better go in the "women's waiting area." I stood at the door way with the door left open. I see about eight men, the 'religious police,' marching angrily toward this young guy. They looked at me, and I think they wanted to hang me! I often wondered if the young guy was flogged for talking to me. I never saw him again.

THE PRINCE

Now we are in Dubai. I was showing the aircraft to the prince of Dubai and his assistant/interpreter explaining the details of the airplane. The Prince did not speak English. Then my boss showed up and took over. I served Arabic coffee. Arabic coffee takes a long time to make. It's very thick and strong. I served it on a silver platter with little stackable cups. As you approach the guest to give the coffee, it is poured in the top cup, the guest takes it, leaving the next empty cup for the next guest, and so forth. The Prince was ready to talk details, as he wanted to purchase of the aircraft. We were all sitting around the table in the aircraft talking about details of the aircraft. I was looking around, not paying attention, when I noticed my boss looking at me. Then he looked at the Prince, then he said, "She doesn't come with the airplane."

Then his assistant approached me secretly, asking if I would be interested in being a private flight attendant for the Prince. The pay was very good, and my housing would be paid for. I had kids waiting for me back home, so I said, "No, thank you."

On the rest of my visits to the Mideast, I always brought my Bible tucked away in my luggage.

A SPECIAL GUEST

My next flight was a domestic flight for a client, David Robinson, who played basketball for the San Antonio Spurs. Based on his prior service as an officer in the United States Navy, Robinson earned the nickname "The Admiral." He was a ten-time NBA all star and a two-time US Olympic Hall of Fame honoree. Sports Illustrated magazine was taking him to Martha's Vineyard for his party, as he was retiring from basketball.

I served breakfast and quickly took their plates away so they could visit. I overheard David sharing Jesus with everyone at Sports Illustrated. They all sat there and intently listening to what he had to say. When we landed,

If I Told You My Story

David invited the crew to the party. It was huge, you had to have paid a huge price to get into this party. We were there three days.

Three days later, we were taking David Robinson back to Los Angeles alone. I served Mr. Robinson his breakfast, and now he had his music playing with ear plugs and was ready to relax for his trip home. After clearing the table and tidying up the airplane, I asked Mr. Robinson as he was seated and looking at his computer, searching for music to relax to if he needed anything else. "No," he said.

He was 7'1" and 235 pounds, which could be very intimidating. I looked down at him and said, "Ya know, we have the same Father." He sat up straight and looked up at me with his face now losing color and asked, "*How so*?!"

I said with a smile, "We have the same Heavenly Father." David looked up at me with the biggest grin that he always has and said, "Oh, yes we do!" Then he invited me to sit and join him.

A Special Guest

David showed me pictures of his family and children. He told me how he was going to start a program to help inner-city children. He told me, "Rhonda, the world just doesn't get it. They ask me, 'Why would you leave basketball when you can make so much more money?'" "They just don't understand. I don't need any more money. I want to serve Jesus Christ and work in ministry."

Then he asked me if I was married. "No," I said. "No one wants me" I told him. David told me, "Oh yes, God has someone for you. You should be married. You go to this church, Phoenix First Assembly, (now known as Dream City Church), and God is going to bring you your husband."

In 2001, Robinson founded and funded the $9 million Carver Academy in San Antonio, a non-profit private school named for George Washington Carver, to provide more opportunities for inner-city children.

Problems were beginning to really rise in the Mideast now. My kids were beginning to not feel so good about me going there anymore. One day, the day before we were to leave for our next trip to the Mideast, Adam phoned me and said, "Rhonda, I'm not going on the trip. Do what you want, but I'm not going." Adam had been reading up on what the government had to say about Americans going into the Mideast. They advised, 'If you're American, don't go. If you're there already, get out.' That's all I needed to hear. If Adam didn't go, I surely wouldn't. That was the end of my career flying private.

HANDS AND FEET

By this time, my son was living in Long Beach, CA and director of an INS company. My daughter met a wonderful Australian young guy. His job would be taking him to Japan for about seven years, so he knew he wanted to take my daughter with him as his wife before the big move. Sherah was on her way to go see Jeff when Jeff phoned me to ask for my daughter's hand in marriage, apologizing for asking over the phone, but under the circumstances, that's how it had to be. That was one of the best days of my life!

I was really alone now, still wondering what I should do if I didn't fly anymore. I searched, and thought I should go into residential real estate. Again, I took time off from flying through vacation time. I took the two-week class to get my real estate license. It was the hardest thing I did. It involved studying 24/7. My house was a wreck.

My pretty little perfectly kept home was now a train wreck with books, shreds of paper, and trash cans that needed emptying everywhere, but I did it! I felt so good about myself. I loved real estate, and I could still fly. My plan was to fly part time and do real estate full time.

At this time in my life, I was alone with my son in California and my daughter in Japan. So it was just me and Sherah's dog, Taylr, the *best* little dog ever! I think she was part human. We watched movies together and ate popcorn. I loved my life. I had started going to church at PFA now. I just loved going to church and being with Jesus people. I made a decision to follow Jesus, and I was going to give 100% of myself to God. There would be no turning back into the world; I was done with that. I realized God didn't need my help to find joy, peace, love, and happiness. It was always there in his outreached hand, waiting for me, just as his word promises.

What will man profit if he gains the whole world and forfeits his soul?
                              Mathew 16:26 NASB

I was in Bible studies and doing as much volunteer work with the church as I was available to do. Every Saturday morning, a group of volunteers from the church would meet at the church, and we would go out in groups to the inner city and knock on doors, delivering bags of food to these people. We would share the joy, hope, and love of Jesus with them if they would be interested. They usually invited us into their homes, and we would pray with them if they wanted to. Tears streaming down their sweet faces as they told their stories.

If I Told You My Story

We loved their children and played with them. They loved it when we would show up. We always invited them to church and told them we had church buses that would come pick them up if they wanted to go to church.

I owned a Black Lexus SC 430 and would drive this car to that area of town. It was my only vehicle, and I wanted to be part of this ministry. God sent angels to protect me and my car. The two years I did this ministry, nothing ever happened to me or my car. This area was bad. Drive by shootings, drug trafficking and more. It could be dangerous.

My usual route was with Becky. We parked in front of the main drug dealer's apartment. Nick had long dark hair with beady brown eyes. A fairly big man. No one was going to bother us around Nick. He loved me and Becky. and always welcomed us into his apartment. Everyone gave us the utmost respect, or they had him to deal with. Nick would always shed a tear and tell us, "I know I need to go to church and get my life back with Jesus." As far as I know, Nick never did.

I was involved with Church on the Street too. We would go down town Phoenix by the railroad tracks and meet with the homeless. A lot of them lived in the area in a box, under a bridge or on a bench. When we showed up we had tables set up with food. Sometimes soup, sandwiches, just lots of food for these precious people. There would be mamas with their children. Young and old men and women coming and telling their stories, never without a tear. We had chairs lined up and a stage. We would all sing and then the Pastor would deliver a heart warming, loving message about how only through God's son, our Lord Jesus Christ, can your life be changed forever and you can be guaranteed eternal life in heaven with Jesus, who sits on his throne on the right side of God the Father. The stage flooded with hungry hearts for love and a change in their life, and they would come to one of us asking for prayer and this life Jesus had to offer.

We helped set up with the right people and counseling to help them get back on their feet and get jobs. Soon PFA/Dream City Church, bought a hotel downtown Phoenix. It became the Phoenix Dream Center. PDC takes in homeless people and gives them a place to live and get on their feet, and of course shares the truth and hope in Jesus. The stories that have come out of this place will bring tears to your eyes. The story that is embedded in my heart is this: It was Christmas. PDC came to our church and spoke about how Jesus changed their lives and what they were most grateful for. One beautiful young girl who was probably no more than eighteen years old got up and told her story. She said she was most grateful for the Dream Center taking her in and introducing Jesus to her and taking her off the streets. She was grateful that her mama could no longer sell her to men for money to put food on the table. Any time I felt sorry for myself, this young girl's story would resonate in my ears.

NO TURNING BACK

I was on fire for Jesus now. I did a lot of spiritual fasting. I did a 45-day fast, 15-day fast, 10-day fast, and 3-day fast. I just loved my Lord God! I had never known Jesus the way I knew him now. How did I miss this?! He is truly my friend! Just like the song says. I have never felt the security and love I found through Jesus Christ. To this day, there is just nothing I don't trust my God for! This is what I had been desiring and searching for my whole life. No more emptiness, only peace. There is nothing like the peace Jesus gives. I thought I would be giving up too much fun! Little did I know the fun I was losing was by not following Jesus!

I loved reading God's word and learning more and more. I could feel the presence of my Savior as I laid out my heart and prayed and surrendered everything. This world is temporary. Billy Graham once said, "The greatest surprise in life to me is the brevity of life." This is so true. Life is such a gift from God. We never know when we will take our last breath so be ready for eternity. We all face an eternity in heaven or hell.

LETTER TO GOD

I had prayed for a husband for a long time now. I even wrote a letter to God with the details of the kind of husband I wanted. I gave the height and color of hair and said that I wanted my husband to love Jesus as much as I did and so much more in my letter. I put this letter under my mattress. When I moved from my condo to my house in North Scottsdale, I lost that endearing letter.

One afternoon, I was in my living room, talking to God. I had my special place in my house that was my prayer closet. I walked around and just talked to God like he was standing in front of me. I believe that is what he wants us to do. I said, "Father, I have prayed for a long time for a husband or to please take the desire out of my heart to want to be married. I said, "Father I believe you have someone for me because I still have this desire to be married. Father, I'm okay not to be married now, I know Heaven is my future. You are my Father, and I know you will take care and provide for me. I also know you want marriage, you made Adam and Eve as husband and wife. I continue to have this desire for a husband, so I believe you have a husband for me. Father, like you did for Isaac and Rebekah at the well, you divinely brought

If I Told You My Story

them together. Father, I'm not doing so well choosing a mate, so I'm asking you as my Father to bring me my husband."

That afternoon after my prayer, I grabbed my Bible and took off for Starbucks. I was sitting at Starbucks outside reading when this guy started talking to me. He was funny but crazy! There was another girl sitting alone with her coffee—a pretty girl with long blonde hair and blue eyes. Soon this guy brought her into our conversation. Luanne joined us. The crazy guy ended up leaving, and Luanne and I became good friends. We were both single. Soon we were trying to think of someone we knew to fix each other up with. I knew of a fun pilot I would fix Luanne up with. I told Luanne, "Jesus is my husband. He's taking care of me." If I'm going to date anyone, my guy had to be a Jesus man, or I wasn't interested. Heaven was my future, I told her. I didn't need a man, I just wanted companionship. Life could be lonely. She said she knew of someone. She said, "I think he even goes to your church. And he's definitely a 'Jesus man!'"

My church was quite large, but I didn't recall seeing a single guy there.

Finally, it clicked. I think I knew who this guy was. Nope, not interested. Yes, he was handsome and looked like the kind of man I would like, but he goes to my church?! What if it didn't work out? Then it would be uncomfortable, and one of us would probably feel obligated to leave.

Nonetheless, we met. It was Fourth of July, and the church always had a celebration. I was polite, but I kept my distance.

Bob was a General Contractor, and I sold real estate. Bob built high-end homes in Paradise Valley. I wanted to learn that market and sell real estate in Paradise Valley, so I talked to Bob. I gave him my business card, and he gave me his. I would see him at church, and we would have small talk. Bob

If I Told You My Story

would walk me to my car after church service. I still was not interested.

I wanted to have a decorative iron door to replace my existing front door. Luanne told me Bob makes those. "He does!?" I began talking to Bob more but I never asked him about the door. A month went by and finally Bob phoned me. He asked if I'd be interested to come see the house he was building in Paradise Valley. Sure! I went to see the home on the side of the mountain. It was stunning!

BOB

Bob was tall, with sandy brown hair and gentle brown eyes that always seemed to smile. He had muscular tanned arms and was wearing his work boots with jeans and a white work shirt that showed off his physique. The sombrero did it for me. He was so cute and handsome!

Bob and I hung out for at least an hour in the hot August afternoon sun. Even though we were inside the house, it was not air conditioned, and it was probably 105 degrees outside. We talked about everything, but our main and favorite subject was Jesus and what an awesome God we served. Bob had such a gentle spirit, and his faith in Christ was real. We could have talked all afternoon, but I needed to let him get back to work, so Bob, the gentleman that he is, walked me down the hill to my car.

When we were almost to my car, Bob, in his soft husky voice, asked if I'd like to go out to dinner sometime. This was on a Friday afternoon. In my younger days, a guy *never* asked a pretty girl out on a date on a Friday afternoon! Don't ya know, she's going to have plans, and the guys are lined up at her door just waiting for an answer. Well, this pretty girl had no guys lined up waiting for her at the door and I'm sitting home with my dog eating popcorn alone watching a movie on TV most Friday nights! I was past that game of pretending I had plans. I said, "Yes."

THE DATE

Bob cleaned his truck the best he could, and came to pick me up in his construction work truck. His truck is his office. He had architectural plans rolled up and tools jammed in the back seat of his truck. In my earlier days of dating, that would have been the number one turnoff. I wanted a nice shiny car coming to pick me up for my first dinner date. My priorities had changed now, and I didn't care about those little incidentals. I used to think I needed a certain look in a guy, and that look included the whole package. What kind of car and what kind of job did he have? What did he have to offer me? That's when I was living in the world

If I Told You My Story

and looked for worldly pleasures and what I thought I needed. God showed me that I had something to offer and bring into a marriage as well. Together, a husband and wife build on that foundation together. Now that my life had changed, the inside of me changed, too. That's just what God does. It's beautiful. The blinders were gone. Of course I still wanted a husband that met some criteria, and I had to be attracted to him, but now I wasn't shallow. I looked at his heart and what he was made of instead of the outside package and what kind of job he had or car he drove.

Wow! He cleaned up even better than I expected. Straight-legged jeans with a soft blue print shirt that showed off his tan. What a gentleman! He came to open the car door for me! I couldn't remember the last time a man did that. That's a lost touch in today's world.

We women created it, though, with women's lib in the 1960s. Women's Liberation Movement brought on a social equality of

The Date

the sexes. Women were not being treated fairly in the workplace. June 1966, while in Washington, D.C., Betty Friedan and twenty-eight women founded the National Organization for Women (NOW). The purpose of the organization was "to take action to bring women into full participation in the mainstream of American society. Exercising all privileges and responsibilities thereof in true equal partnership with men." I remember when this began. I was twelve years old going on thirteen, in the seventh grade, and in study hall with my girlfriends, looking at a magazine on Women's Liberation. It showed women going braless. We laughed and said we would never do that.

Now, I'm not saying this movement didn't need to happen. There definitely needed to be a change, but it got to where women were offended if a man opened the door for them. Women needed then and now to be recognized for their work and receive the same wages as or even better wages than a man, if their work merited it. Men changed with the movement as well, and chivalry became a thing of the past. Bob was not one of those men. He liked treating a lady like a lady even if he was not equal to her status, and I liked that about him.

We went to Pure Sushi at DC Ranch, a trendy restaurant with an inviting ambiance for sushi lovers. DC Ranch Crossing is a casual neighborhood center that includes restaurants and a variety of distinctive shops that's fresh and contemporary. It embodies the atmosphere of a European mountain village with lush courtyards and fountains. This was a very romantic atmosphere in which to walk around the village and window shop. A perfect first date. The date went well, and Bob was smitten. There definitely would be a second date.

We started seeing each other weekly and then twice a week, with phone calls in between.

The Date

If a few days went by without hearing from Bob, I missed him. This man really made me feel like a princess. A girl should always feel like she is the most beautiful woman when she is with her man, and I was that to Bob.

When Bob walked me to my door after a date, he started to get brave and would always bend down to give me a kiss goodnight. I would always turn my cheek. Bob was everything I was looking for in a man, and I loved being with him. He made me feel like nothing would ever harm me again, because he would be there to protect me. We laughed and enjoyed being together. What was wrong with me?!

Finally one evening after walking me in to my house, as he was getting ready to leave, he said, "Every time I try to kiss you, you turn your cheek. I'm beginning to like you. If you're not interested, I can go away." My heart started pounding. I didn't want him to go away! That got my attention. I prayed and asked God to help me with this. This man had everything I needed and desired in a man to be my husband, so why was I holding back? I told him that it just takes me a while, but I was enjoying his company.

Bob's birthday was September 14th. I suggested we drive up to Jerome for the day for his birthday. I thought maybe the drive and doing something out of town would change things and I would relax. I was going to treat him to lunch in Jerome for his birthday. I wanted to share that beautiful little ghost town that was thriving with tourism with him. It's one of my favorite places to go even to this day, and it's only about a two-hour drive outside of Phoenix.

We put the top down on my Lexus convertible, and away we went. We took the back roads as opposed to the freeway and left early in the morning to meet the crisp morning air in the cool pines after a quick stop at the coffee shop, of course. What a drive for a

The Date

beautiful day. Driving through the winding mountainous roads in a convertible was like being on a magic carpet ride. Bob didn't say anything, but being the kind of man that he is, taking care of the woman, I'm sure he would have liked being the driver on these winding roads as opposed to being the passenger.

After walking around the town of Jerome and up and down hills, we worked up an appetite. With the aroma of burgers, soon our noses took us to a popular restaurant for lunch. After lunch, we walked to the copper mining museum. Lots of interesting facts to make falling in love with Jerome even easier. Why couldn't I fall in love with this man, though? Bob drove us home, and I was glad to let him.

PAYSON

A week went by and we didn't see much of each other. Bob is a patient man. Bob suggested us going to Payson for the day. Payson is another two-hour drive north east of Phoenix in the mountains. Its location is very near to the geographic center of Arizona. Payson has been called "The Heart of Arizona. Surrounded by the Tonto National Forest which opens up many outside activities.

In 1918, author Zane Grey made his first trip to the area surrounding Payson. Zane Grey wanted to be a baseball player, but settled as an American dentist and author best known for his popular adventure novels and stories associated with the Western genre. He built and lived in the one-room cabin for a time. Grey wrote numerous books about the area and also filmed some movies, such as *To the Last Man*, in the Payson area in the 1920s.

I would bring my kids to Payson in the summer when they were little and camp out at Christopher Creek. We slept either in the car or on a sleeping bag outside the car. All three of us snuggled up together. We visited Zane Grey's museum many times before it mysteriously burned down.

Now there's a new one, a replica, and it's open to the public. This one is safe from forest fires and natural erosion because it's located next to a lake in the heart of Payson. Sometimes we drove up to Payson alone, and sometimes with friends, but we were always sure to have a great time. There's just nothing like cooking breakfast out in the open. Once again, God had to have sent Angels out to protect us from the wild animals and intruders.

If I Told You My Story

Bob was in Payson by himself for a three day fast earlier in the year. Just him his Bible and a jug of water. He wanted to take me where he stayed on his fast deep in the Coconino National Forest on the Mogollon Rim where the Tonto Apache Indians once hunted. It was beautiful! Trees, clouds and sky overlooking the picturesque lakes.

The Mogollon Rim is a rugged escarpment that extends across the entire forest and provides excellent views of the forest and the valley.

The smell of the pines and the beautiful wild flowers is breathtaking. It was as if the pines were talking to us as the wind gently blew the trees with a sound of rush between the trees.

I brought a picnic basket filled with sandwiches, fruit and sweets, *and* my Bible. We sat on the cliff overlooking the vast mountains in the silence and cool mountain air. There was no sound but that of the wind brushing up against the trees and our breathing. We read our favorite scripture after lunch. It was so serene and peaceful that it was a bit awkward, because I still did not have a spark of a romantic interest igniting a flame in my heart for this wonderful man. When we get home I have got to tell Bob this was not going to work out. I didn't want to waste his time or hurt him. I'm still struggling with emotions. What is wrong with me?! I'll never find a man as perfect for me as this man, but I can't fake it, and he's too wonderful! He deserved better.

Our drive home was the same. We chatted about our day and how fortunate we were to live in such a beautiful state. Bob wanted to stop at an antique store on the way home just outside of Payson. We were just about to leave Payson, and I looked up at Bob, and all of a sudden it was as if God took me by the shoulders and shook sense into my head! It was as though magic love dust was sprinkled all over me. It was as though God was saying to me, "My daughter! You asked me as your Father to find a husband for you. A husband that would love and adore you and take care of you. I just hand-picked this man for you and put this man in your life, and you're about to send him away and ruin my plan for you!" As I was looking up at Bob, I fell in love that very moment! I am as serious as a heart attack! It was magical! That has never happened to me in my life! I had such a brick wall up that I wouldn't allow the gift God was giving me to penetrate through my wall!

Honestly, I think I was probably in love with this man at the beginning and kept trying to find an excuse not to allow it to happen therefore putting up my brick wall. Bob could tell something had changed. I looked up at his gentle face and smiled, and then allowed him hold my hand. He squeezed my hand

and smiled. We stopped at the antique shop. Before getting out of the car, he pulled me close to him, gazing into my eyes he touched my face with his fingers and caressed my hair and then bent down and softly kissed my mouth. Then he told me, "Anything you want in here, I'll get it for you." Oh my! Lucky for him, I didn't find anything I had to have.

GOING TO THE CHAPEL AND WE'RE GONNA GET MARRIED

Bob dropped me off at home and and asked if I wanted to get dinner. "Yes!" Now I never wanted to be away from him. He picked me up a couple hours later, and we went to get pizza. That night he took me home, and I wanted him to stay awhile. We watched some TV until it was time for him to go. Now I wanted to taste his kiss and accepted it without turning my cheek. Bob told me he loved me as he kissed me good night. Whoa! It was going a little too fast now. I just decided I liked you! We became inseparable. It wasn't very long before I became brave enough to return the sentiment: "I love you too."

It was right before Halloween. Our church was having the "Trunk or Treat" for the children on Wednesday night as opposed to the usual Wednesday night service, so we left and went to dinner. Love felt so good. We were so enamored with each other. This was real love. It wasn't an infatuation like other relationships in the past had been. Bob cared about my every being and it showed, I felt his love. We were always laughing at silly things. We never wanted our evenings to end. Bob had to work the next day. I was off and didn't have to fly. We hated our good byes. Bob kissed me good-bye, and the words just came out of his mouth. It wasn't planned, and I think he surprised himself when he asked, "Will you marry me?"

I didn't have to hesitate when I answered, "Yes!"

THE WORD IS OUT

It wasn't long before all my flight attendant friends found out Rhonda Bond was getting married. I'd been a flight attendant for over thirty years, so I had a lot of friends. At this point in my life, I was fine without a husband. I certainly didn't "need" a husband; I wanted a husband. All my friends knew this. No one even knew I was dating anyone. So when they heard I was getting married, I got phone calls and emails asking, "Is this true?!" It was all over Facebook. What is that last name? Bush-*what?* 'Bushwty.'

Bob had been married before. He had two kids, Jonathan and Emily, who were both teenagers when we married. My son was living in California and my daughter was married and living in Japan. I sent them both an email that I was getting married. Blown away I got a phone call from them both almost at the same time. "What?! You're getting married and you sent it in an email?!" They, too, didn't know I was dating anyone. My kids knew this man had to be special. This wasn't like their mom.

When I realized a guy didn't love Jesus the way I did, we would be unequally yoked and it would never work out. My priority for a husband now was he had to be a 100% organic believer and follower of Christ, a Jesus man. I told both my kids we thought about flying to Hawaii to get married. They both said, "No, we want to see you get married."

"You do?" I truly didn't think they would care what I did. So I planned my wedding on that response. I got married at the amazingly, beautiful, Wrigley Mansion in Phoenix, Arizona.

The wedding and planning the wedding was just amazing! God was in this from the very beginning. It was becoming my little girl dream come true. Luanne went with me to look for a dress. I found my dress at the first shop we went to. We wanted to honor God's commandment, so we planned our wedding for January 17, 2009. A quick courtship of only three months.

Because of the timing of our wedding in January we were able to get a great deal at Wrigley Mansion. The wedding planner kept expanding rooms for my wedding. Can I just describe my little girl dream wedding?

A DREAM COME TRUE

I truly felt like a young girl getting married. I felt like a princess!

Everything turned out perfect! We had a brunch wedding. The reception was stunning!

I ordered a three tier red velvet cake with white cream frosting decorated with fresh red roses. Each table was just beautiful! White linen table tops with a small vase of roses. Gold charger place settings. Wedding favors on each table was a silver candle in the shape of Cinderella's pumpkin carriage, representing me, the Cinderella. I knew my friends would understand that gift. There were also silver-plated luggage tags in the shape of an airplane.

The wedding planner gave me a bridal room and a groom's room to get ready in with spectacular views overlooking the city. She She catered our rooms with coffee, water and juice, bagels and fruit. My room was enchanting.

An old but charming 1932 style, very feminine, overlooking the city with a mountain view. Jeff and Sherah joined me with the kids in my room and along with my best friends, and of course my sis and hairdresser, Brenda.

The harpist played, and my friend, David, sang "You Lift Me Up." A white runner led to the wedding platform from the stairs. My grandkids, Riley and Cade, were in the wedding. Riley was so excited the day of the wedding

that she went into Cade's bedroom at 4 o'clock in the morning to wake him. "Cade, wake up, we're going to Nana's ball today." Sherah and Jeff could hear it on the speaker in their bedroom.

Four-year-old Riley dropping red rose petals from her basket on the white runner.

A Dream Come True

Now here comes three-year-old Cade behind her, all serious and walking down the white runway with my wedding ring tied to the pillow. Bob was supposed to have taken my ring and place it in his pocket, but he forgot, so now Bob and the Pastor were hoping Cade had my ring on the pillow he was carefully carrying down the runway. Just adorable, these children were. Giggles and *ahhhs* can be heard from up the stairs, where I was standing behind the interior wall.

I'm alone in my room. I'm too happy to cry so I pray. The wedding planner comes to get me. "It's time." The wedding is outside and the

weather is perfect: 78 degrees and cool. White wooden chairs were placed on the grassy lawn. Two flights of stairs led me down to meet my guests and husband-to-be. My handsome son met me at the top of the stairs with the biggest smile. I appear and my guests all stand and turn to look at me and my son holding my arm and carefully walking me down the stairs. I first passed my friend, Luann, and

A Dream Come True

she smiled and whispered how beautiful I looked. That's all I needed to hear to feel confident now to meet my man. I think my son was as happy as his mama. I sure was happy he was there to support me and turn me over to Bob.

As I'm walking down the stairs my Bob looked up at me as if he was saying, "that's my bride and she's so beautiful." The wedding couldn't have been more beautiful! Bob and I said our vows, and when we walked past our guests, we had the biggest smiles of love and happiness. Most importantly, Jesus was there. I could feel his presence and he proved once more to me that he's a "detail God."

But seek ye first the kingdom of God, and his righteousness; and all these things shall be added unto you.

—Matthew 6:33 (NKJV)

We have three beautiful grandchildren. Riley, Cade, and Brynlee. These babies truly make Nana's heart smile. They are the love of my life!

What I find most endearing about my husband is the fact that every single morning, Bob gets up an hour before leaving for work to spend time with the Lord reading his Bible and in prayer.

I sold Bob's house, and he moved into my house. It wasn't long and Bob was ready for a bigger house with a garage he could work in. Bob loves to work on old cars as a hobby, plus he has his decorative iron business. I found us a home in Paradise Valley. We had just moved into our new home, and everything was still in boxes.

Bob and I are sitting in the office when a letter fell to the floor. Seriously, it seemed like it fell from nowhere. I picked it up. It was the letter I had written to God describing the husband I wanted. At the end of my letter, I added, "Prepare me to be a Proverbs 31 woman for my husband-to-be." Tears of joy filled the room.

I showed it to Bob, and he said, "Wow! this really is me!"

And that's what God does; he gives us the desires of our heart.

Delight thyself also in the Lord; and he shall give thee the desires of thine heart.
—Psalms 37:4 (KJV)

CPSIA information can be obtained
at www.ICGtesting.com
Printed in the USA
LVHW090851070719
623341LV00001B/1/P